Georgius Forbes

Alternating and interrupted electric Currents

Being based upon three Lectures delivered to the Members of the Royal

Institution

Georgius Forbes

Alternating and interrupted electric Currents
Being based upon three Lectures delivered to the Members of the Royal Institution

ISBN/EAN: 9783337069254

Printed in Europe, USA, Canada, Australia, Japan

Cover: Foto ©ninafisch / pixelio.de

More available books at **www.hansebooks.com**

ALTERNATING AND INTERRUPTED ELECTRIC CURRENTS.

BY

PROF. G. FORBES, M.A., F.R.S., F.R.S.E.,

ETC.

BEING

Based upon Three Lectures delivered to the Members of the Royal Institution.

ILLUSTRATED.

BIGGS AND CO., 139-140, SALISBURY COURT, FLEET STREET, E.C.

PREFACE.

The present book is the outcome of a course of three lectures delivered before the Royal Institution of Great Britain in 1895. The verbatim report of these has been revised and corrected, and forms the substance of the book. The lectures were primarily intended for an intelligent audience, but one not well versed in electrical science; hence, technicalities have, in general, been avoided, and the colloquial form of expression has been retained in the book. The attention of telegraphists may, however, be drawn to the mechanical torsion model of a submarine cable—described in the first and second lectures—as illustrating in a novel and remarkably accurate manner the various phenomena observed in working with an actual cable. It is believed, too, that many of the other mechanical analogies described, which formed a chief feature of the course, will be found of interest to others besides.

My special thanks are due to Mr. H. E. Mitchell for his assistance in preparing these lectures for the press.

GEORGE FORBES.

34, Great George-street, Westminster, S.W.,
20th December, 1895.

CONTENTS

	PAGE.
INTRODUCTORY	9
ANALOGIES TO ELECTRIC PROPAGATION	11
OBVIOUS AND NON-OBVIOUS FACTS	13
FARADAY'S DISCHARGE EXPERIMENT	19
ANALOGUE OF AN ELECTRIC CIRCUIT	19
CAPACITY	21
MECHANICAL ANALOGIES	25
SUBMARINE CABLES	28
WORKING TORSION MODEL OF CABLE	35
DEDUCTIONS FROM MECHANICAL ANALOGIES	38
ELECTRICAL UNITS AND THEIR MECHANICAL ANALOGIES COMPARED	41
OERSTED'S DISCOVERY	43
FARADAY AND INDUCTION	43
SELF-INDUCTION AND INERTIA	44
MUTUAL INDUCTION	50
DANGERS OF SUDDENLY BREAKING CERTAIN CIRCUITS	55
TRANSFORMER ON OPEN CIRCUIT	60
PRACTICAL UTILITY OF ANALOGIES	70
FURTHER ALTERNATING-CURRENT PHENOMENA	71
INDUCTION WITH HIGH FREQUENCY	73
SKIN RESISTANCE	79
ELECTRICAL RESONANCE	88
ELECTRICAL OSCILLATIONS	92
EXPERIMENT BY LORD ARMSTRONG	95

ALTERNATING AND INTERRUPTED ELECTRIC CURRENTS.

▼▼▼▼▼▼▼▼▼▼

INTRODUCTORY.

When I was asked by Sir Frederick Bramwell to give a short course of lectures here, I enquired of him what subject he would like me to lecture on, and he said, "Some physical subject." I am aware that at the Royal Institution it is generally the custom to deal with principles rather than applications of science.

Now it happened that in the course of the work upon which I have been engaged during the last few years I have had occasion to try and find some simple means of expression to convey to my own mind, and also to the minds of others, certain somewhat obscure electrical phenomena which take place especially when using alternating or interrupted currents, and I arrived at a method of producing mechanical analogies to the various cases of the propagation of electricity which was of such. very great assistance to myself, and which enabled me so much the more easily to convey some intelligent idea of these curious electrical actions to others, that I thought it might be well to incorporate them in the present lecture.

Prof. Dewar told me yesterday that you would be expecting me to talk about the Niagara Falls, and that you would be disappointed if I did not. I can only say that I cannot imagine why any such opinions should have been held, and I can only express regret if I am disappointing those present by taking up a different subject. However, I have this to say, that the present lecture does deal with some of the work which has been going on there, while it deals far more directly with submarine telegraphy. The next two will deal almost entirely with problems which have been cropping up in the work of utilising those great falls and distributing the power in the neighbourhood, so that I trust the course will not be altogether a disappointment to those who had expected a different one.

Before beginning my subject I have also to say that I have a good many experiments to show you to-day, and in the following lectures, which I do not wish you to think for one moment are original—many are somewhat striking, but they have been evolved after very patient work by others than myself. I particularly mention the name of Mr. Campbell Swinton, who has placed at my disposal some extremely beautiful apparatus which I shall have frequent occasion to refer to. Others have assisted me much. Dr. Muirhead has been invaluable to me in the assistance he has given me in illustrating the present lecture by enabling me to produce before you some of the phenomena which occur in submarine cables. This is a subject which has hitherto been wrapped up in a certain amount of obscurity—of mathematical symbols; and while dealing with that it is my hope that I may be able to divest the treatment of the subject of those difficulties, and make the phenomena which appear in submarine cables generally intelligible to anyone. This I hope to do by means of a working model which I devised, and which Dr. Muirhead has constructed for me.

Analogies to Electric Propagation.

When we speak about the propagation of electric currents through an Atlantic cable, we know that while the electricity is passing into the cable a charge of electricity is being imparted to the outside of it—the cable is being charged, and that prevents the rapid propagation of the current through it. Now, it is all very well to talk about it in that kind of language, but it is not very intelligible, and for this reason, that it is difficult to understand. We do not understand exactly what we mean by electricity. All the time we are using such a phrase, we are trying to grasp some mechanical analogy which shall make it intelligible to us.

For instance, in dealing with the propagation of electricity, it has been very common in the past to treat of the subject by the analogy of a flow of water through pipes, and an excellent analogy it is. The pressure of the water is the electromotive force which compels the water to move, the velocity of the water acquired in the pipe is the current, and the friction of the pipe represents the resistance of the wire. If you have a pipe in a complete circle, with a centrifugal pump at one part, the power being expended in overcoming the friction of the pipe, you have there the analogue of a battery with a certain pressure, on a closed circuit carrying a current, with a certain amount of resistance; where the velocity of the water increases and increases until the backward force caused by the friction of the pipe is equal to the motive force, or the pressure of the water, in the centrifugal pump. That is the usual analogy employed for the most ordinary electrical phenomena.

We might extend that analogy, and instead of using up the energy of the centrifugal pump in overcoming the friction of the pipe uniformly along its length, which is equivalent to heating the conductor, we might place a very narrow piece of pipe at one part of it. The energy would then be almost entirely consumed in that narrow pipe at the far end. Now, that is analogous to the incandescent lamp circuit of this room. The centrifugal pump is the dynamo in the central station, the pipes that go round are the wires that go to and from the central station, the narrow piece of pipe at the far end where the power is used up is the analogue of the incandescent lamps. Instead of placing that small pipe at the end, we might have a little water-motor, such as a turbine, and we should then have the central pump, a limited amount of whose power is used up in friction in the pipe, driving the motor, which might be doing useful work. There we have the source of power in a central station, in which the centrifugal pump represents the dynamo, the pipe representing the wires along which a current is flowing with a certain amount of resistance offered, and the turbine at the far end representing the electric motor which is driving the machinery.

That is a very useful and simple sort of analogy, but when we come to some more obscure points this analogy fails to be quite so easily understood, though in many cases it is a very good one indeed. I am speaking about the difference between ordinary cases which are easily understood, and cases which are somewhat more obscure. In the ordinary ways of electric lighting in a house or from a central station the method of the propagation of electricity is comparatively simple, and any person who has played with a battery and wires is able to see at a glance what will happen when you connect the wires to the dynamo and then to the lamps. But there are other phenomena which come in later, especially when using alternating currents, which are not so easily understood.

Obvious and Non-obvious Mechanical Facts.

As I am going to adopt mechanical analogies to show the electrical phenomena, I may well here adopt a mechanical analogy to show you the different class of phenomena which I wish to illustrate. There are many mechanical phenomena which are perfectly obvious to any intelligent person when looking at them. If you have a steelyard with a heavy weight supported at the short arm and a light weight at the long arm, it seems perfectly natural that the heavy weight may be supported by the light weight, without going into the mathematics of it or measuring the quantities and proportions; the result is a very obvious one to every person. When I exert pressure upon a screw press, I, with a very little pressure and a large motion of my hand, produce a very small downward motion, but a huge pressure upon a copying-book, for instance: that is a transformation of mechanical energy which is perfectly obvious. And so with the different mechanical powers—the lever, the wheel and axle, the screw—any of these are perfectly obvious.

I will now show you one or two phenomena not quite so obvious, and the first which will occur to everyone of you here is simply the gyrostat, which everyone of you know, but which I wish to show you because it illustrates what I am working at (Fig. 1). All I wish to say with respect to the gyrostat is that it takes some time and some thought to realise why it is that the wheel does not twist in the direction in which the obvious forces are impelling it. That is a thing requiring a considerable amount of thought to enable you to arrive at a correct conclusion. Why does it twist round a vertical axis, and in which direction does it twist? These are things which are not perfectly obvious, and which do require a considerable amount of thought to make them clear. If I were able to show you a mechanical analogy which would make them clear, it would be a great step. In the

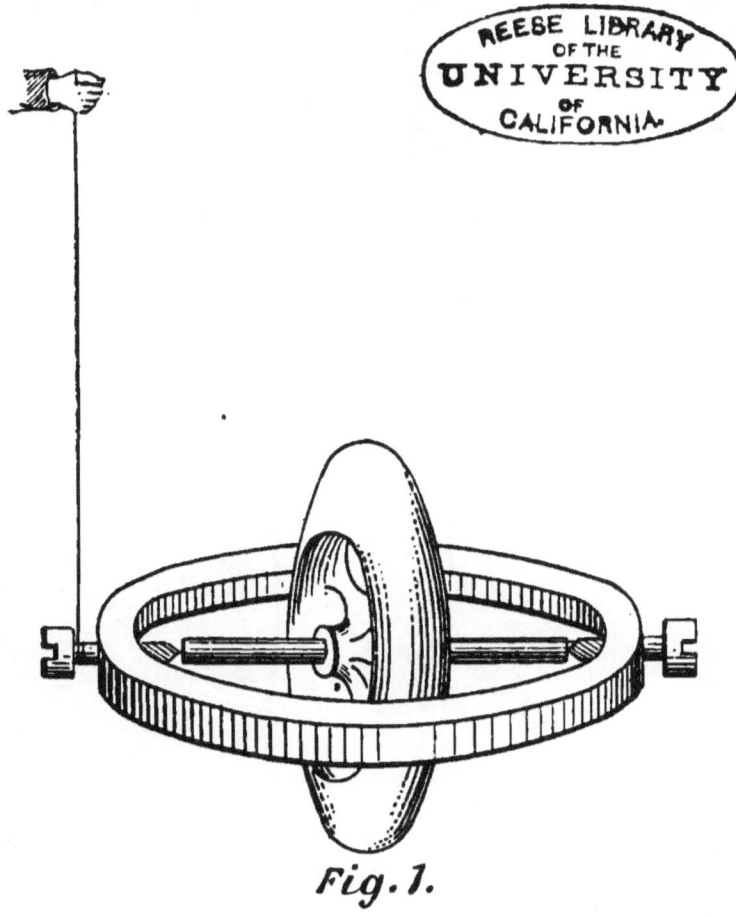

Fig. 1.

case of the electrical phenomena I think I shall be able to show you some analogies which are very complete.

16 ALTERNATING AND INTERRUPTED

Here, again (Fig. 2), I have a chain caused to revolve by means of an electric motor, where the motions which take place are those which are not perfectly obvious even to the very well-trained

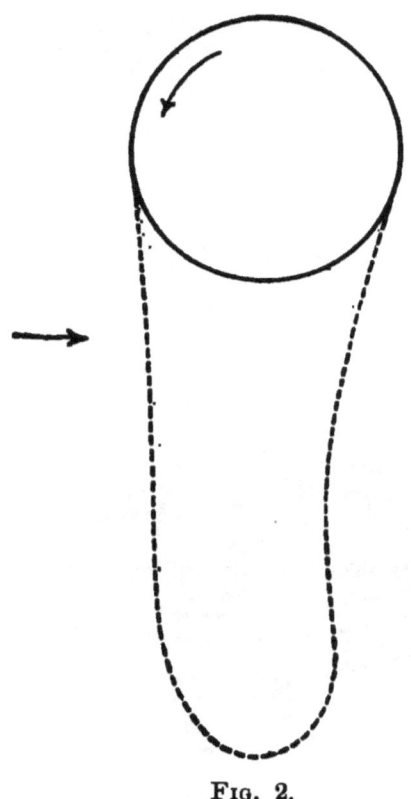

FIG. 2.

mechanic. [*Experiment.*] When I draw this chain towards me it acts somewhat like a rigid bar. If I impress a motion on it, it retains that motion; and if I draw the chain out for a considerable distance and let it go, it goes through a series of motions which are somewhat difficult to understand, which

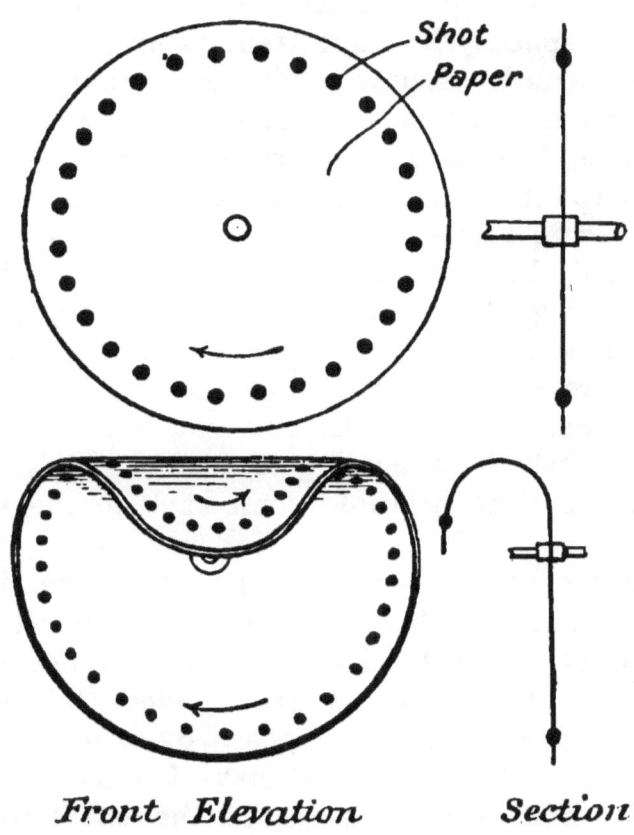

Front Elevation   Section

Fig. 3.

you would not have foreseen if you had been asked to say what the action was to be.

I will not delay you with these any longer; a

series of beautiful experiments may be done with them. I will show you another example, however, where the ordinary intelligent man with mechanical knowledge does not immediately realise what is going to happen when a certain experiment is going to be performed. I have a disc of ordinary cartridge paper loaded at the circumference by means of weights, and I shall give a considerable speed of rotation to the disc, and I shall bend this disc while it is in rotation, and I do not think that the result will be exactly as everyone would have expected (Fig. 3). Now we leave it alone, and I do not think that everybody here would have expected it would retain the same shape continuously for so long a time. I go faster, and you see that there is a force which is tending to lead it back to the vertical. In time it would come back to the vertical position.

These are all phenomena dependent upon inertia, and a great number of those in the electrical experiments which I wish to show you are equally dependent upon what is called electrical inertia. Electrical inertia usually goes by the name of "self-induction," and I shall have frequent occasion to use it; and as you have seen the effect in these cases, you will remember inertia is the analogue of self-induction. These experiments also involved a certain pliability (*e.g.*, of the chains), flexibility, ability to move, and a natural tendency to return if not affected by other causes.

Many of the electrical experiments I shall have to show you depend for their action upon electrical flexibility or yielding, which usually goes by the name of electrical capacity. These two things, capacity and self-induction, are the two most interesting things which come into the course of these lectures. I shall have to deal with them pretty largely.

Faraday's Discharge Experiment.

An experiment of Faraday's may be mentioned as a fitting opening to this course of lectures. He discharged a large number of Leyden jars, and tried to discharge them through a lengthy wire; but the discharge preferred to cross from one end of the wire to the other end instead of going round the circuit, although that metallic conducting circuit was there, and you would have thought it would go round it. In that experiment is involved nearly the whole of the principles of the various things I wish to bring before you in the course of these lectures. Instead of bringing all things grouped together in one experiment, and then separating them out, we shall now proceed to take them separately.

Analogue of an Electric Circuit.

I wish to show you before we go on what is the mechanical analogue which I wish in these lectures to follow up in the study of an electric circuit. The water analogy which I have spoken of is very useful, but I am anxious to draw your attention to another mode of illustrating it, and I will just do so by means of a model here. I have an electric circuit represented here, by means of this piece of apparatus (Fig. 4). My hands represent a central station at the top, and there is a factory at the bottom where the power is going to be used in lifting weights. These are suspended on the vertical indiarubber ring I hold. I have to apply an electromotive force. The force is purely a mechanical one—a twisting force. I then give a certain twist. By applying the force up here I apply a certain twist to the indiarubber tube, which is transmitted

to the bottom, and the weights rise. That is a very simple thing, and therefore it is all the better for illustrating what I wish to say. We may use a continuous current, in which case I always wind up in the same direction, or we may use an alternating current, when I twist in opposite directions. The resistance in this case is represented by the distortion of this indiarubber, but any other form of friction would do just as well. For example, if I take one of these strips of telegraphic paper and twist it round at the top, there is a certain amount of air friction resisting the twisting; or I might suspend in water or oil anything I wish to twist, when the friction between the surface of the object and the fluid would exemplify electrical resistance.

Capacity.

Now, one of the first things I wish to give a little idea about is capacity. Of course, as I mentioned, an Atlantic cable has a certain amount of capacity, a considerable capacity, and it shows itself by a charge upon the outside of the cable. To illustrate this more clearly, I will just go through the experiment of charging a Leyden jar and then discharging it, just to show you what I mean by capacity and what I mean by something which will hold a charge. [*Experiment.*] Now, an Atlantic cable consists of a wire surrounded by insulating material. The wire is the one conductor—we will call it the internal coat—and the sea is the outer one, and this is to be charged to the potential or pressure which you are using. Now, to charge any portion of it, involves the passage of a certain quantity of electricity through the cable. But the current cannot pass instantaneously through the whole length of the wire, because the condenser action is demanding a certain supply for every length of cable as you go from one side of the Atlantic to the other, and the resistance of the wire will not allow it to flow more rapidly than at a certain rate with the number of batteries you happen to be using. That is the foundation of the difficulty that arises in dealing with submarine cables.

I wish to illustrate this question of capacity to make it a little more clear, and will do so at the risk, perhaps, of confusing you a little; but I will try to avoid that. I have here an arrangement by means of which I can charge and discharge a condenser with extreme rapidity. The amount of electricity which a Leyden jar can hold is extremely small. Faraday used to say that in decomposing a drop of rain there is far more electricity used than there is in a lightning flash.

The little Leyden jar which I used just now is capable of holding but a very small quantity of electricity, but the condenser which I am going to make use of immediately is capable of holding still less. It holds very little electricity indeed. But by using a current interrupted with extreme rapidity, and reversing it, I am able to supply the current at very rapid intervals, so that the total quantity which passes in one discharge is very small, but the total quantity which passes in the course of a second is very large indeed owing to the enormous number of discharges which take place in the course of a second. In the experiment which I am going to show you, by using two pairs of tin plates we have two condensers of far less capacity than that of the Leyden jar which you saw just now, and yet their capacity will be sufficient with the enormously high pressure I am going to use, and the very rapid alternations, to provide a current sufficient to light an incandescent lamp, which is really a considerable quantity. I now take the two top plates of the condensers and place them over the others, and you see a light glowing (Fig. 5). I can place them still lower, and you get a brilliant illumination even using that extremely low capacity, simply owing to the hundreds and thousands of volts we are using and the enormous frequency of alternation.

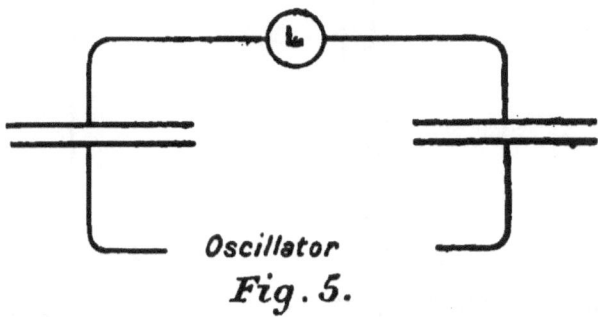

Fig. 5.

While I have this apparatus I will show you the effect in a more exaggerated manner. I am going to use simply two points separated by a glass plate. The capacity of this arrangement is something almost infinitesimal, and yet by using these enormous electric pressures and the high frequency which we are employing, we are able to get electricity enough to light a lamp. I lean a glass plate against one of the terminals, and to the other terminal I attach an incandescent lamp with a wire leading up to the glass plate. Now, the capacity of that condenser is really almost infinitesimal, and yet there is a sensible amount of electricity passing into and out of it during the

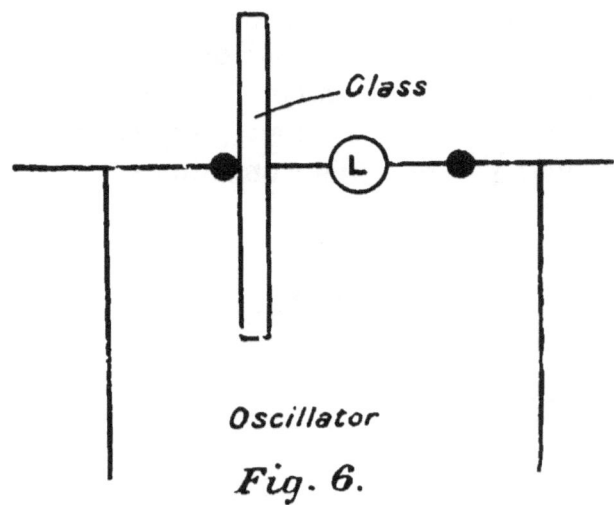

Oscillator

Fig. 6.

time I am exhibiting it (Fig. 6). This is a very striking experiment, and as I have got the apparatus here I will show you what the capacity of

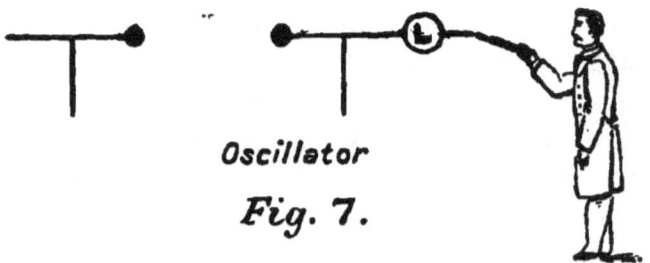

Oscillator

Fig. 7.

my own body is. I think you may take it that this ground is completely insulated, and that there

is no current capable of going through me to it.
Therefore, since the lamp is only attached to
one of the terminals, the current has no complete circuit to pass round, and when I touch this
it is only by charging my body and then discharging my body that the lamp is able to light up
(Fig. 7). There is the lamp lighted, the charging
and discharging of my body being utterly imperceptible to myself.

Mechanical Analogies.

Now, capacity is what I am going to deal with
chiefly to-day, and most of the rest of the lecture
will be devoted to illustrating an Atlantic cable.
But I will first show you the kinds of models that
I have found instructive, and they are of the rudest
and most elementary character; and as soon as
you realise what you can do by thinking over these
simple models, and the perfect analogies you can
get to the electrical phenomena, I am sure it will
be a very great help to you in working out any
electrical problems or in thinking about what is
going on in connection with any.

Well, in the first place, I am dealing altogether
with twisting of materials. I consider that the
electromotive force is the force with which I twist
them. I can measure the force with which I twist
anything by winding a wire round it and letting it
go over a pulley and putting a weight on it (Figs. 8,
9, 10, and 11). That force is the electromotive
force. Here (Fig. 8) I have a central station with

a twisting force, here I have the line going to a distant place, and there I have a resistance, and a resistance eventually such as to equalise the force I am exerting. There is also a definite velocity of rotation attending, which is analogous to the current. Therefore, in these models, twisting force

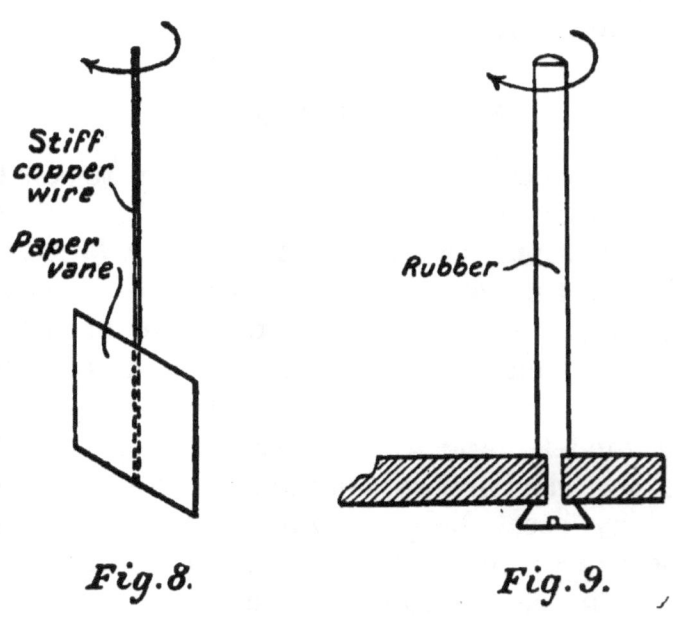

Fig. 8.   *Fig. 9.*

is the electromotive force, friction is resistance, and velocity of rotation is electric current. In such a station I can have either a continuous current, ever going round in the same direction, or I can have an alternating current, going first in one direction and then in the other. In either case, if the wire is perfectly stiff, it is pretty easy to follow the phenomena which are going on.

I have a number of other examples. You may take any mechanical devices for these illustrations, and

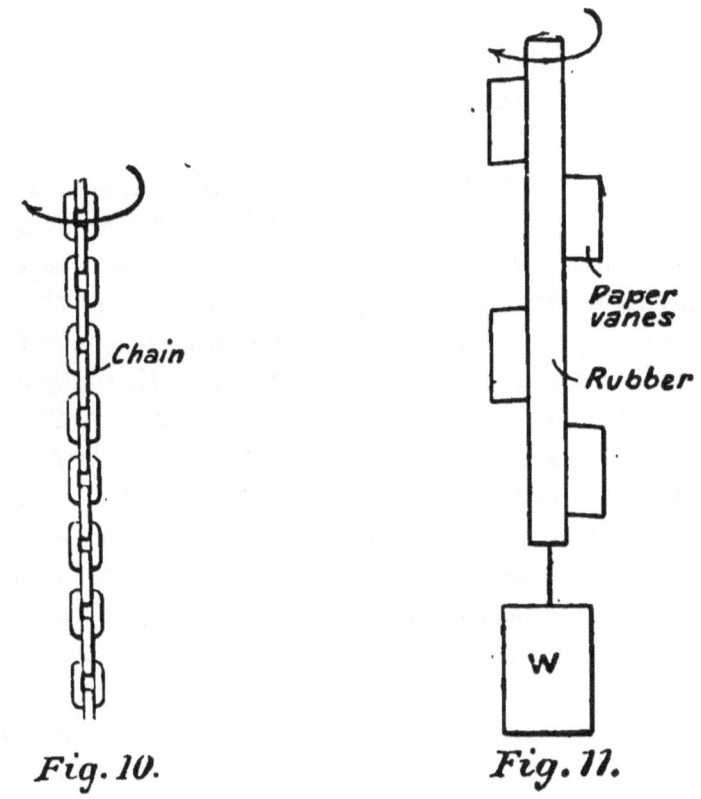

Fig. 10. Fig. 11.

you can find electrical analogues. Here is a case (Fig. 9). Suppose I twist that elastic string at the top and keep it fixed at the bottom—what is that equivalent to? It is equivalent to a cable of large capacity which is insulated from the earth, and not free. In that case there is a certain charge given to it with any electromotive force—in other words, there is a certain twisting given to it, with an ounce weight, for example. Therefore we are able to conceive of capacity as a yielding in the same direction, and the charge is the amount of that yielding. A twisted, flexible thing like this is a charged cable. I have here (Fig. 11) something which represents a combination of flexibility in the cable, which is capacity, and of mass or inertia at the far end, which is self-induction, and of air friction, which is resistance. If I pass alternating currents through that, you will notice that the progress is in a sort of wavy motion down the length of the line. Exactly the same thing happens in the electrical case, provided you have your units in the two cases proportionate to each other.

Here (Fig. 10) I have a heavy chain. When I twist this chain round backwards and forwards it does not all rotate at the same time; that is due partly to its having pliability, which is capacity, and also inertia, which is self-induction. The inertia of the mass prevents the motion being communicated to the whole length at the same time.

Submarine Cables.

And now I wish to draw your attention to submarine cables. I will first simply show you how a deep, narrow trough of water is an analogy of a submarine cable. This trough has capacity at every part of it because the level can rise, it has resistance all along because there is friction between the water and the sides; in every way the propagation of a wave along the trough is analogous to the propagation of a wave in a submarine cable—in fact, it is a very accurate analogy indeed. To illustrate it, I will simply make a few waves, and you will see the propagation of the waves along the whole length of the water surface, the change of level taking some time to reach the far end. That is what happens in the case of a submarine cable—it takes some time to reach the far end.

I wish now to show you a most curious thing. I have here a thread, suspended from a spring, with a set of vanes attached, and two nozzles for blowing in either direction against the vanes. The thread is immersed in the glycerine and water in this long tube, and at the bottom I have a mirror attached, from which I can reflect a beam of light. The thread might be attached to the bottom of the tube by a spring, but instead of having a spring I have used the obvious method of attaching a small magnet to the mirror, and using controlling magnets. Let me refer you to this diagram (Fig. 12). The top spring acts as a condenser, and the airblast and the vanes take the place of the battery. The spring below takes the place of the other condenser. There we have mechanical apparatus taking the place of electrical as used in sending messages through an Atlantic cable. But before using it I wish to show you the phenomenon of the retardation of the current in an Atlantic cable—that is to say, by means of those artificial cables which are made by Dr. Muirhead, who has assisted me so much. Dr. Muirhead's artificial cables are made

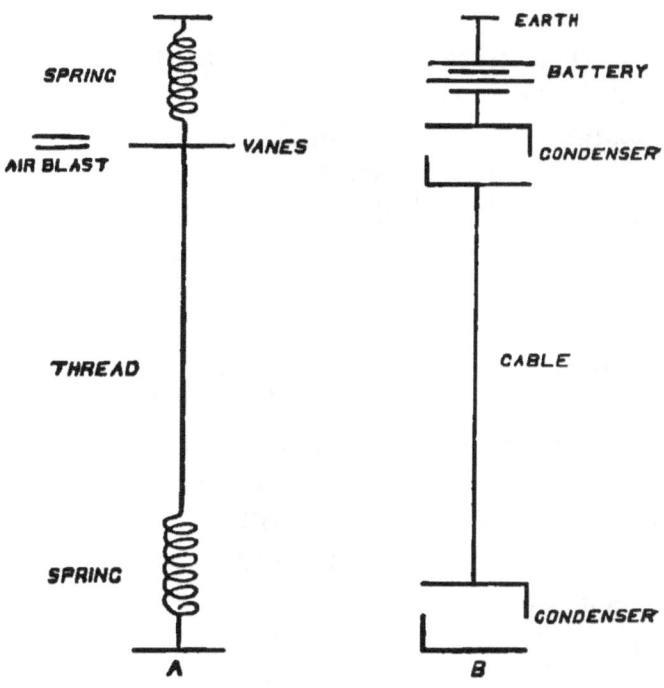

FIG. 12.—A B represent Earth.

of tinfoil, cut in these shapes (Fig. 13) so as to have high resistance and a proportionate capacity. These two kinds of plates are placed alternately side by side and separated by paraffin, the plain sheets being connected with the earth. He builds up his cable that way, and thus makes an Atlantic cable. Now, to show the retardation of

ELECTRIC CURRENTS. 31

current. This galvanometer is connected with the sending end; the current will then pass through the whole length of the cable and come out at the other end, and, passing through the galvanometer there, deflect it. The movements of these galvanometers will be seen by spots of light shining on the screen. Now, Mr. Edgar is going to send

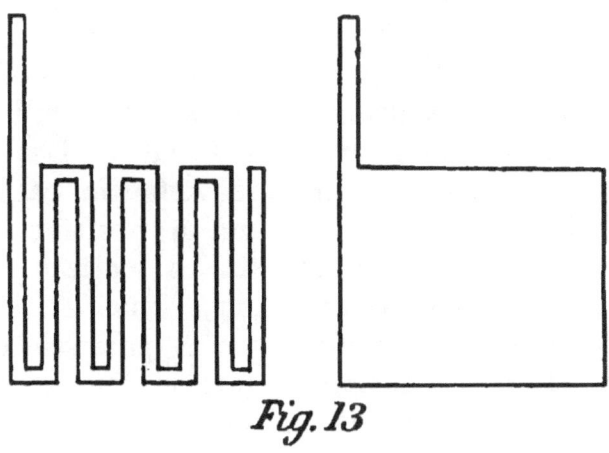

Fig. 13

a signal through the cable, and what you will have to observe is that the sending galvanometer is deflected immediately, and that it takes a sensible time before the other is affected. [*Experiment.*] You can see the interval of time between the sending signal and the arriving signal, and you also see how gradually the arriving signal creeps up. Exactly the same is noticed when I rotate the vanes at the top of the thread hanging in the tube of glycerine that I have just described to you, and the deflection of the mirror at the bottom is noted. This, I have to say, is really, if you choose your units, a perfect representation of the Atlantic cable. I would like to devote a little more time to show how, by measuring the twist, you can get the capacity and the resistance. Time will not permit me to do that, but I may perhaps have more to say about that on another occasion.

I will now show you some further illustrations with these. I will begin by using the suspended thread in order that we may see and foretell what we ought to expect from the cable. We will use the suspended thread first without a curb. Let me tell you what a curb is. In the Atlantic cable it is found that if you simply send a positive current through, the signal arrives in a certain manner which is perfectly satisfactory, but it is much improved if, after having given a positive current, you give it a short time afterwards a negative current. I will put it in another way. If you take a whip with a long lash, and you want to make a crack you may give a forward motion to the lash, and you get a certain amount of cracking; but you get a much sharper crack if you follow it with a backward motion. So here, curbing with my cable means giving it first a twist in one direction and then a twist in the other. Not curbing simply means giving a twist in one direction only. We will signal the letter S—that means three dots—to show you what takes place in a real cable. You will follow it by means of the spot of light you see on the screen. I want

ELECTRIC CURRENTS.

to show you the difference between curbing and not curbing. When you are telegraphing quickly, the signals bank up on the top of each other, and cause the spot of light to travel further and further away as the second and third dots come. With curbing this does not occur. Now, by depressing first the positive key and then the negative we can produce the equivalent of curbing, and Mr. Edgar will be able to show you that by means of the reflection on the screen. [*Experiment.*]*

In conclusion, I wish to draw your attention to the beautiful apparatus which is used in the operation of transmitting through an Atlantic cable. In the first place, the message is punched out on a piece of paper, and it is punched so as to send the current in either a positive or negative direction, according as the holes punched are on one side or other of the strip. The strip of paper is afterwards passed through the transmitter, and as it is passed through contact is made through the holes, and the message is sent. We will now transmit a message through this. [*Experiment.*] The messages which are sent through the cable are received on an instrument called a syphon-recorder, which was originally invented by Lord Kelvin. There is in the curb sender I have on the table a most beautiful device, for while it is working I can, by turning this screw, vary the amount of curbing in signalling through the cable, and by varying it you are able to adjust the signals so as to get the most beautiful

---

* See note at end of lecture for more detailed description of the model.

trace of the recorder possible; and also by this means, not using a condenser, but by direct signalling, you are able to get an increase of speed in

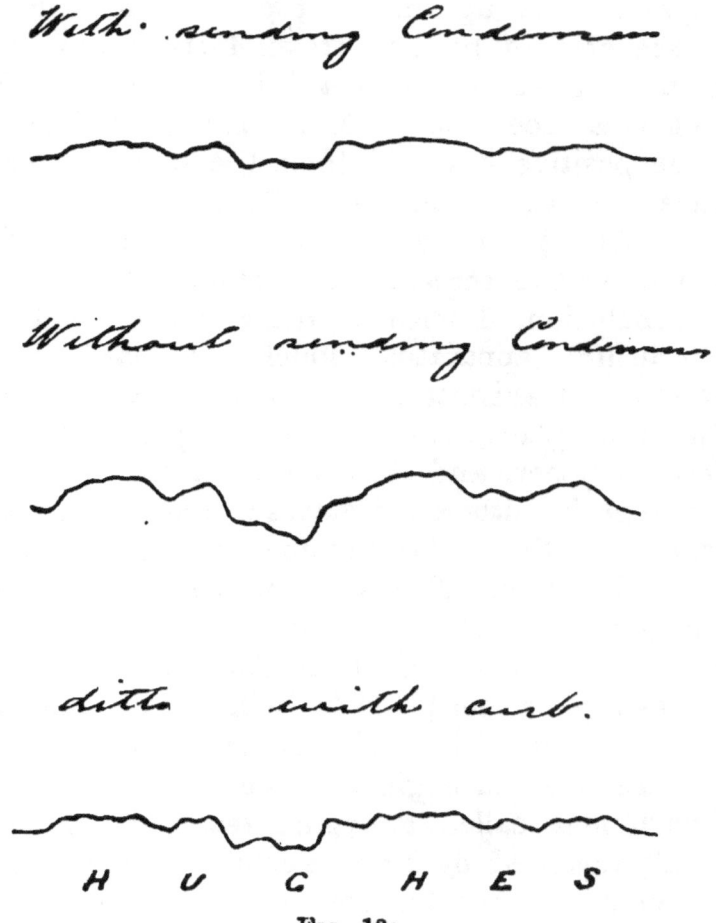

Fig. 13a.

some cables of about 25 per cent., which is a matter of great importance. I do not think there is time

ELECTRIC CURRENTS. 35

to go into much more, but I am able to show you a few specimens on cards of some signals made in three different ways: first, with a condenser; second, without a condenser; and third, without a condenser and with a curb, and the third is the best. I wanted a word with a good many special letters like "h" and "s" in it, and so I selected the name of a member of this institution who has had a great deal to do with telegraphs, and there (Fig. 13A) are the marks exactly as they came out drawn by the recorder itself—"Hughes."

NOTE.

Working Torsion Model of Submarine Cable.

For the benefit of telegraphists, I wish to say a few more words about the model illustrating submarine cables.

If you hang a thread in a vertical tube filled with glycerine and water in suitable proportions, you have all the essentials for understanding the propagation of electric pressure through a submarine cable. But the measurement of the twisting force applied at the top is difficult. Any mechanism having mass or inertia is objectionable, as this introduces the analogue of a heavy self-induction, which does not exist in the electrical problem. For this reason, I suspend the thread, and support on it vanes, which can be blown round by a constant pressure of air to represent the electromotive force. The spindle of the vanes is solid, and goes down into the liquid to the point where the thread representing the cable is attached. The vanes and air-jet represent the battery, and I find it convenient to have two jets, to rotate the vanes in opposite directions. These are actuated by two keys, + and −, which admit the air to one or other of the nozzles. If there be no sending condenser to be represented in the model, the vanes must be supported by a silk fibre without torsion. Of course the vane spindle is held laterally by bearings. Things being so arranged, we may assume that the air pressure used represents the electromotive force of the sending battery, say 100 volts. If the cable which we wish to imitate has 800 microfarads capacity, the permanent charge on it would be $\frac{800}{10^6} \times 100 = 0.08$ coulomb. Fixing now the lower end of the thread, apply the air-jet and observe the twist. Suppose it is ten revolutions. Then one revolution means 0·008 coulomb, and one revolutions per second means 0·008 ampere. If now the resistance of the cable be 6,000 ohms, the permanent current would be $\frac{1}{60}$ ampere. This should be represented by $\frac{1}{60 \times 0.008} = \frac{1}{0.48} =$ two revolutions per second. Leave now the lower end free, and apply the air-jet at the top. If you get two revolutions per second, you have the right resistance in the model. If not, add more water or glycerine till you get it right.

You cannot, however, get mirror signals thus, as

velocity indicates current, and the zero would move. If, however, you put a mirror at the lower end of the thread, and restrain its motion either by a spring or by magnets, you are putting in a receiving condenser, and the deflection indicates the potential at that point of the cable, just as a recorder does. You ought to add here, below the mirror, a vertical wire to represent the resistance of the recorder. If now, instead of having a torsionless fibre at the top, you attach the vanes to a spring, this is equivalent to a sending condenser, and you have a perfect representation of the signals received by the special cable which you have imitated.

LECTURE II.

In my last lecture I indicated my desire during the course to illustrate a good many of the phenomena that occurred in using interrupted and alternating electric currents by means of mechanical analogies; and I think you will have already seen that, in certain cases at least, these mechanical analogies are of very great help to us in understanding what is going on. I propose to carry on the same method—through the present lecture at any rate. I concluded the last lecture by showing you by means of such mechanical analogies that we were able to form a clearer conception, as it seemed to me, of what is going on in a submarine cable than we would be able to without such mechanical analogies.

Deductions from Mechanical Analogies.

Now, I am very anxious to prevent anyone here from being led to draw too much from these mechanical analogies, and I warn you that a considerable amount of care is required in dealing with them. In going a step further, sometimes one is apt to miss the complete analogy, and so to be led to wrong deductions. This is not the case with the submarine cable which I showed you. In that case the analogy is, I may say, almost absolutely complete. There is only one defect in that analogy, and it is a trivial one—the resistance to the current is represented by the friction between the glycerine and water in this tall tube, the thread inside which is being twisted, and the current is represented by the velocity of rotation of that thread. Now, the force opposing the motion in the case of this glycerine varies as the square of the velocity. In the analogy the resistance ought to vary as the velocity. The opposing force varies as the square of the current instead of as the current, and that is the only defective part of the analogy; and it is really of very little consequence. As you all saw last day, it does not interfere with the very perfect representation of what is going on in a submarine cable. In fact, those who are accustomed to read the signals for a submarine cable agree that they cannot be distinguished from the signals given by this twisted thread. I said last lecture that these analogies would enable me to dispense with mathematical formulæ in explaining some obscure phenomena to you, but I would also wish to say that I think that these analogies are of more use really to the mathematician than to anybody else, because such analogies enable the mathematician to gain a clearer grasp of the quantities he is dealing with, and to treat them more easily than he would be able to do otherwise. In fact, I may say that Clerk Maxwell, in the whole of his great work on electricity, was merely working with mechanical analogies, although he does not say so in the book. It is obvious to anyone who reads that book that he was working out mechanical problems the whole time, knowing that the mechanical results at which he arrived would apply equally well to the electrical phenomena, and that instead of putting mass, momentum, and all those mechanical quantities, he put the electrical equivalents. That is perfectly obvious, and if any proof of it was wanted it is in his invention of that term, " electric displacement," which is a little puzzling to the electricians; but as soon as we realise that it is a mechanical equivalent the difficulty disappears.*

Now, I was a little hurried last week at the end of my lecture in the demonstration of the apparatus I have just referred to, and it is such a very complete reproduction of what goes on in an Atlantic cable that I wish to make it a little more clear. At the top of this thread I have a spindle, with guides or bearings, the whole being supported by a spring. On the spindle is a disc with vanes. There are also two brass jets, through either of which I can send a current of air to twist the thread in one direction or the other. These

---

* As an example of the importance of mechanical analogies to the philosopher, see the remarks by Prof. Larmor, F.R.S., in his paper, "A Dynamical Theory of the Electric and Luminiferous Medium," *Phil. Trans.*, 1894, Part II., p. 719: "The credit of applying with success the pure analytical method of energy to the elucidation of optical phenomena belongs to MacCullagh ; he was, however, unable to discover a mechanical illustration such as would bring home to the mind by analogy the properties of his medium, and so his theory has fallen rather into neglect from supposed incompatibility with the ordinary manifestations of energy as exemplified in material structures."

currents of air are manipulated from below. I may blow into the pipe, or I may use a reservoir of air, as I did on the last occasion, and the rotation is given to right or left by the two keys on the table, which open the air-valves and let the air pass in. The string itself is so chosen that it represents a long Atlantic cable. The way in which the measurements are made is this. To see that it represents a cable, you first, with a definite pressure of air, allow the string to rotate at a steady speed. Suppose it makes a revolution in a second. Then, with the same pressure, holding the bottom part of the string, or restraining it by means of magnets, apply the same air pressure and see how much twist it gets. It gets, perhaps, ten revolutions. Divide that ten revolutions by the one revolution, and that ten gives you a constant of the cable; it gives you the ratio of the charge to the current, or the product of the capacity into the resistance ; that defines the quality of the cable exactly, and the number of turns that the top of the string can acquire defines the capacity of the cable as a condenser. I will not delay you further to explain the ways in which this is an extremely close analogy, because it will become tedious. Those of you who have followed me so far, and have any interest in submarine cables, will see the very close analogy which there is.

Electrical Units and their Mechanical Analogies Compared.

I have placed in the accompanying table a number

of electrical units which can be taken, and their analogies. I have simply tabulated them as a means of reference. I have given examples of certain units which may be assumed; for instance, volt and ampere. From these all the other units will follow. I have assumed 1 volt to mean 1 ounce driving at a radius of 1 inch, and 1 ampere one revolution per second. A coulomb is equivalent to one revolution; there is a farad when it makes one revolution with 1 ounce of pressure; a henry, if you have a current of 1 ampere gradually reduced during one second of time by its own self-induction to zero.

⸺Electric⸺		⸺Mechanical⸺	
Potential.	Volt.	Twisting force.	1oz. at 1in. radius.
Charge.	Coulomb.	Twist.	1 revolution.
Capacity.	Farad.	Twist per unit force.	1 revolution with 1oz. at 1in.
Current.	Ampere.	Rotation per second.	1 revolution per second.
Self-induction.	Henry.	Moment of inertia.	62oz. at 1in. radius.

The pressure that is in that interval of time being constantly exerted is the equivalent of 1 volt—and that is obtained by the momentum of a cylinder of 1 inch radius weighing 62 ounces. That is all I need say about that at the present time. Of course, these units might be varied, and when we are dealing with submarine cables we will take very different ones from these. On the other hand, we might take, for instance, an ampere to be three hundred revolutions per second. So, when we are dealing with transmission of power from one place to another, if we wanted to make an analogy we would use other units from these. In each particular case we select our own units.

Oersted's Discovery and Possible Deductions Therefrom.

The primary discovery in electromagnetics was made by Oersted in 1820, when he showed that an electric current would deflect a magnet. From that simple experiment the philosopher might have retired to his chamber and worked out the whole of the electromagnetic phenomena which we are at present acquainted with. He could have seen that the north pole tends to rotate round the wire carrying a current in one direction and the south in the other. He would have foreseen Faraday's experiment, in which a pole was made to rotate; he would have seen that in producing this mechanical action some power was used, and therefore the current was diminished and a counter electromotive force set up in the wires; he would have seen that a wire acted as a magnet, and would consequently, when properly wound round a core, induce magnetism in a piece of soft iron.

Faraday and Induction.

But of all these different phenomena, the one I wish to direct your attention to most to-day is the work of Faraday. I am going to show it in its crudest and most simple and elementary form. I have here (Fig. 14) a single wire, which I shall connect with a battery through which I can suddenly send an electric current. In its neighbourhood, but not touching it, I have another wire which is not connected with any battery, but is connected with the galvanometer at the far end of the room. What I wish to show you is that at the moment of making an electrical contact in the one wire, a current in the opposite direction is generated in the second one, and on breaking it a current in the same direction as the primary current is

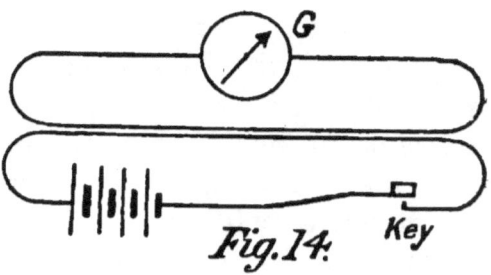

Fig. 14.

created; that is to say, we ought to get a deflection of the galvanometer connected with the second wire in one direction when we make the circuit, and in the other direction when we break it. Remember the galvanometer shows a current generated in the wire which has no battery connected with it.

Self-Induction and Inertia.

This phenomenon of induction is somewhat equivalent to that of inertia; it may be considered as its equivalent in its effect on the current in the primary wire. It sets up a counter electromotive force in the wire through which the electric current is sent. It is just like the effect produced when we set a mass in motion—there is at first an opposition to its motion, and that is due to its inertia, and is a good analogue of the induced current.

Now, you will see we have a spot of light upon the screen perfectly steady. I will send a positive current in one direction through one wire, and the spot of light, which indicates the current which goes through the other wire, is deflected to the right. I will now break the circuit, and the spot of light goes in the opposite direction, showing that the current is induced in the opposite direction when we break the circuit. Now, a certain amount of power is being generated in the second wire, which is, of course, being taken out of the primary wire, through which the current is going—that is to say, momentarily an electromotive force is acting against the current in the primary wire. That I have said is analogous to the effect of inertia. For example, here (Fig. 19) I have a rod with a heavy flywheel at the base of it. If I try to give a rotation to it with my finger there is at first an opposition to the motion—there is a sensible opposition, and it takes some little time to acquire the steady rotary motion which it eventually obtains. That resistance which my finger experiences is the counter electromotive force. But once I have started the thing, and it has acquired a steady rotation, there is no such counter electromotive force acting at all. If I try to stop it I feel the effect of the flywheel trying to pull my finger round, and trying to continue its motion in the same direction. That gives us a clear idea of the self-induction that is taking place.

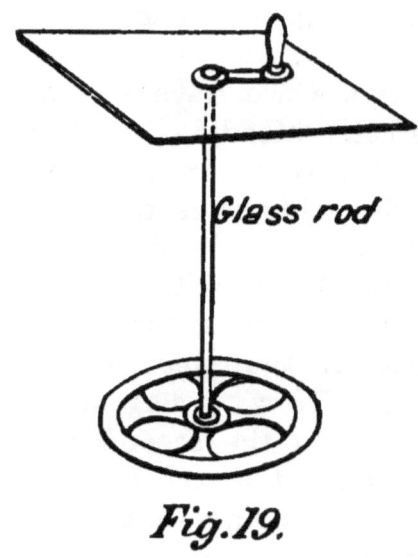

Fig. 19.

Application of Foregoing Experiments to Engineering Work.

The application of these experiments and phenomena really belongs to great engineering works. In these we have frequently to do with alternating currents instead of continuous, and you will be able to see, by the analogy, that we have the induction of the current producing a very remarkable difference in its effects in the two cases. After I have once got the flywheel started by applying a very slight force at the top, the flywheel has nothing to do with the rotation. But now suppose I try to produce an alternating current with the same force backwards and forwards. I find that I am not able to get up the same velocity, that the inertia of the flywheel interferes most materially with the freedom of rotation, and I am not able to get the same speed or the same current through my circuit by that means as I did with the continuous current.

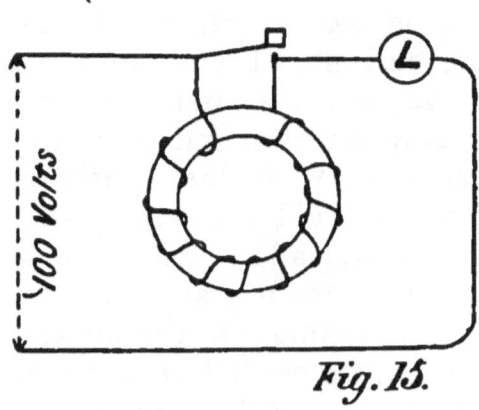

Fig. 15.

That that is the case, I am going to show you very simply by means of a lamp, which is a very good indicator of the strength of a current. I have here a coil of wire, which is really a Gramme ring, and I will pass a current of electricity through this coil of wire and then through the lamp (Fig. 15). First, I will use a continuous current, and show you that the lamp at once acquires its full brightness. I will then use an alternating current, and show that the behaviour is very different, because here in the self-induction of this coil we have the inertia of the current acting in the same way as the inertia of the flywheel that I showed you. We will now try a continuous current. Now the current is simply passing through the coil of wire and through the lamp, and being a continuous current, there is no self-induction and no counter electromotive force generated. Consequently, when I short-circuit the coil you see no preceptible increase in the light of the lamp. The resistance of that coil is very slight indeed, and produces no effect. I will now put on an alternating current. I cut out the self-induction coil, and I have now the alternating current passing through the lamp directly, and you see it gives the same brightness as with the continuous current—using 100 volts as I did before. I will now cut out this short-circuit and allow the current not only to pass through the lamp, but also to pass through the self-induction coil acting as the flywheel to which we are trying to give oscillating movements, and the lamp is immediately greatly diminished in brilliancy, as the effect of the reduction in the current produced by the coil.

One of the greatest helps given to electrical engineering by the discoveries of Faraday has been the invention of the induction coil, or, as it is sometimes now called, the transformer. This can only be used with alternating currents. I have a specimen of a transformer on the table here. It consists of a core, of a bundle of iron wires, through the centre, with a coil of wire wound round it; around that again there is a separate coil of wire, and presently I shall show you that when we change the strength of the current in the one wire a current is generated in the other wire. In fact, the transformer gives really a magnified effect of the induction of two wires upon each other. That we were only able to detect by means of the galvanometer. The action of the transformer is much more powerful. The iron core is introduced, and instead of having a single wire, a number of coils are wound round, and both increase the effect to a great extent. The transformer has been and is of much use to electric light engineers, because it enables us to convert a current of one pressure into a current of another pressure. If we have a large number of turns of wire on one of these coils, and a small number on the other, we may convert a high-pressure current into a low-pressure current, or *vice versâ*. That has been of great use in electric light engineering, enabling us to carry a high-pressure current to great distances over a thin wire, and to transform it at the point of consumption, where the users of electric light desire to transform the current to one of low pressure, such as is suitable for electric lamps, and perfectly safe to handle.

But if that has been of use in electric light engineering, it has been infinitely more useful in electric power distribution. When I first had to investigate the problem of transmitting energy from a power station at the Niagara Falls to different consumers who might be near at hand or far away, the first thing that struck me was the enormous difference in the character of the electric currents that different people wanted. Some people wanted it for lighting, some wanted it for power, some wanted to generate electric currents for electrolytic operations; some wanted continuous currents and some wanted alternating; some wanted 100 volts and some 50 volts. I do not say that it would have been impossible to have supplied these demands by means of a continuous current, but there would have been very great difficulty in doing so, and in every single case it would have been necessary to introduce a motor driving a dynamo. The expense and the trouble of such a distribution would have been simply enormous. And although some very high authorities — I may mention Lord Kelvin among others, he almost alone now—persistently said that the continuous current would have been the best for the purpose, I am glad to say that in that great work the alternating current has been used, and it is a matter for congratulation to everybody concerned that the further we proceed with the work the more we see its adaptability to all the different circumstances that arise.

Mutual Induction.

Now, with regard to this mutual induction which takes place in the transformer, I am going to show you the effect in a more striking way than by using the alternating current. I am going to show you that, with the continuous current, when you break the one circuit you are able to get a very powerful electromotive force in the other circuit. I shall pass a continuous current at 100 volts through the fine coil of this transformer (Fig. 16); I shall then break the circuit suddenly, and you will be able to see that, at the moment of breaking, an induced current is created in the secondary coil which will

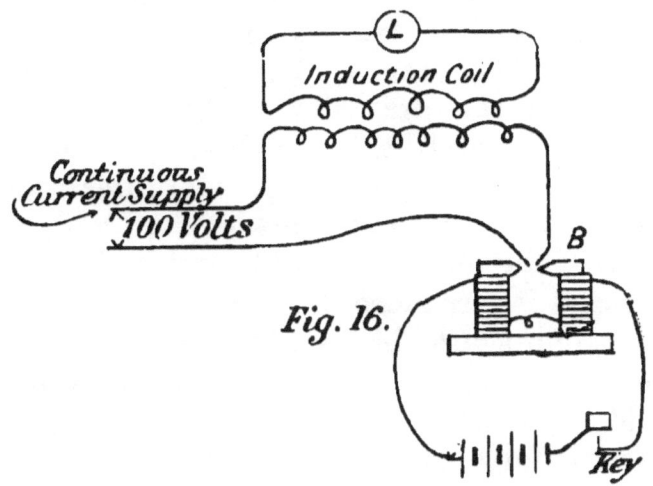

Fig. 16.

be sufficient to light up the lamp in circuit with it. I will first show you this without producing any effect—that is, with the electromagnet unexcited. I have now closed the primary circuit, and when I break the circuit it will not be sufficient to light the lamp. The induced current is not sufficient to light that lamp, although I break this circuit as sharply as possible. The reason is that an arc is formed, as you perceive, where I break the circuit.

This is equivalent to putting in a graduated resistance which prevents a sudden break to the current—causes the break in the circuit to be gradual and slow; whereas it is the suddenness of the break that causes the illumination of the lamp.

I will now excite the electromagnet, and you will see that when I break the circuit at a point between the poles of the electromagnet, the electromagnet will not allow an arc of light to remain. The tendency is for the current of electricity to be deflected. That arc of light is the current of electricity, and the arc will be blown out almost instantaneously, and this enables me to get a sudden break instead of the gradual break which I had before. Now I have made the contact, and as soon as I break it you will notice that a much sharper spark is produced than before, and that the lamp now lights up brilliantly; and by having a very powerful magnet it is possible to produce a very sudden break indeed, and consequently a high electromotive force in the secondary circuit.

These different phenomena of induction, pleasing as they are, and useful as they are to the electrical engineer, are matters that require considerable caution in their handling. There are some troubles which would arise if sufficient attention were not paid to them, and I am going to deal—it may seem at an unnecessary length—with some of the analogies to show you what happens when we suddenly break a circuit possessing much self-induction.

ELECTRIC CURRENTS. 53

Clerk Maxwell's Model Illustrating Mutual Induction.

Before doing that, I wish to show you a model, which I am enabled to do by the kindness of Prof. J. J. Thomson, which was introduced by Clerk Maxwell to illustrate this phenomenon that I have been describing to you—the phenomenon of induced currents in a secondary wire or coil. It is an instructive experiment to look at, and it is an interesting one, because it shows how thoroughly Clerk Maxwell's ideas were connected with mechanical phenomena when he was working out his electrical theories.

Here you will notice that I have two wheels (Fig. 17)—one which I am turning, and a second one which is not turning at all. The one which I am turning at this moment represents the primary circuit, and the one which you see is not turning represents the secondary circuit. These wheels are connected, not by a rigid axis, but by means of three bevelled wheels in the centre—one keyed on each of the two horizontal shafts, and the third running loosely on a shaft, at right angles to the others, forming one of four cross-arms, having heavy weights at their extremities, and therefore possessing considerable inertia, and so resisting any tendency to set them in motion round the horizontal shaft of the hand-wheel, to a sleeve running on which they are attached. The consequence is that, if I suddenly try to turn this handle round, the inertia of the intermediate gearing will prevent it from turning round at first, and the secondary wheel will turn in the opposite direction to that in which I am turning the first—*i.e.*, if I do it suddenly enough.

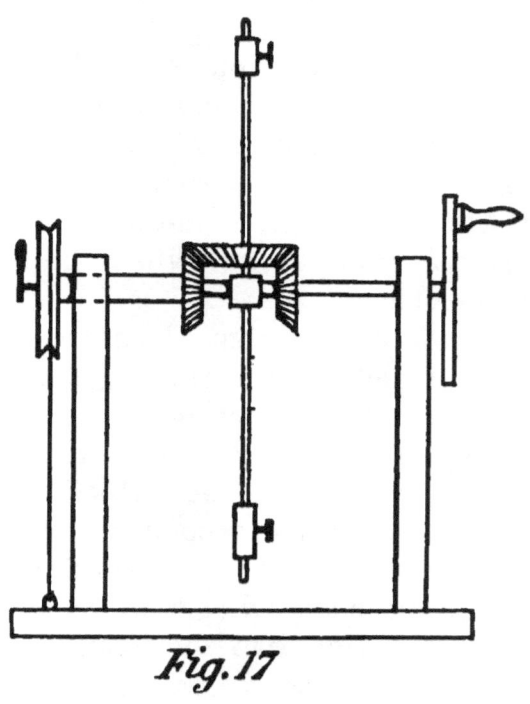

Fig. 17

But now the intermediate gearing is rotating, and if I suddenly stop the hand-wheel, then the secondary circuit goes on rotating in the same direction in which my hand has been going. This is a very beautiful analogy, but, of course, you must not look upon it as more than a rough sort of analogy. It is not an explanation.   it were an explanation we should have to say, I suppose, that this beautiful gearing and these masses at the end of the cross-arms are parts of the ether. In the true case a motion is given to the ether in the one direction, and the moment the ether has motion it transmits it to the other wire. That it could be an explanation is, of course, quite out of the question. But a mathematician finds a very great help from getting a mechanical thing to illustrate what he wants to arrive at in his theories. I remember that in the celebrated Baltimore lectures which Lord Kelvin gave in 1884, he was trying to work out a theory of an ether and vibrating molecule giving light, and he worked out then and subsequently a view of the ether in which every part of the ether consisted of an assembly of gyrostats. I remember perfectly well his saying : " I do not believe ether is made up of gyrostats, but I am never contented with any theory until I can get a mechanical analogy to satisfy me. After we have the mechanical analogy which satisfies the conditions, we can then work on to get something in the ether which will be analogous to it in the method, though different to it in the mechanical conception."

Dangers of Suddenly Breaking Circuits Possessing Heavy Self-Induction.

I want now to show you some of the analogies which help to show us the enormous dangers which may be incurred in practice by suddenly breaking our circuits which have very large self-induction. In an alternating-current system like that which has just been put in at Niagara Falls, the self-induction of the transformers is something very large indeed. And if we were to permit insulated cables to be laid in the earth over considerable distances, the capacity would also be very considerable of these cables. The combined effects of the capacity and of the self-induction would lead to a great deal of trouble, and it has been my effort to reduce the capacity of the cables, and to lay them in such a way that there should be as little capacity as possible, so that that source of trouble should be as much obliterated as could be. Also I introduced the plan of lowering the frequency of the alternations as low as possible, which also assists largely to get over troubles due to capacity. I need not go into that question of the reduction of the frequency at the present moment; but I may say that the further we have gone in the matter the more thankful we are that we realised the necessity in time, and have lowered the frequency to a figure that has never been reached before. It is not only an enormous safeguard to us in countless other ways, but it enables us to use apparatus of a simplicity which would have been impossible with the higher frequencies.

Now, neglecting the question of the capacity, let us deal simply with the question of the self-induction. It is remarkable what great difficulty I have had in getting people to realise the danger that there is in suddenly breaking these circuits containing enormously powerful self-inductions; but I have insisted throughout that it is indispensable to the success of that great work at Niagara that no circuit in which these powerful effects of self-induction are operative shall ever be suddenly broken. I found manufacturers relying simply upon their past experience, and not looking into the new conditions caused by the enormous size of the machinery. I found able manufacturers unable to realise this point. They had noticed, when they broke the circuit of an alternating current with some self-induction, large sparks at the switch. Several had the notion that, if they could get rid of these, that was all they had to do. Their desire was to make the break at the switch as sudden as they possibly could, so as to prevent sparks and a train of fire from appearing there. As a matter of fact, by doing so they were endangering the whole system by introducing an enormous electromotive force of self-induction tending to break down the whole system in one part or the other. I will show you one or two mechanical analogies to illustrate this; and, first, I will adopt the analogy of flowing water which I spoke of last time, which is an extremely useful analogy.

I have here (Fig. 18) a pipe through which I am able to allow water to flow downwards. There is in the middle of the suspended part an opening, which is at present closed. A little brass tube projects from the main tube, and that brass tube is closed simply by means of a membrane of paper, which prevents the water from escaping there.

E

58   ALTERNATING AND INTERRUPTED

That piece of paper in this analogy represents the insulation on the cable, and what we are going to test just now is the dielectric strength—that is to say, the breaking strain of the paper. If that

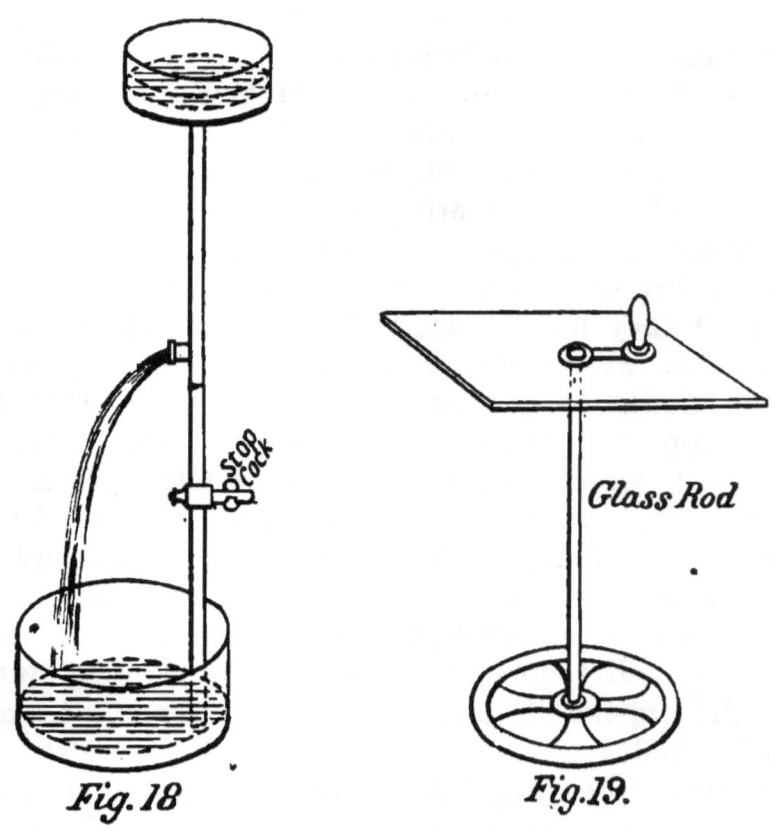

Fig. 18.   Fig. 19.

paper gives out when we suddenly stop the water, then the analogy is, that when we stop the electric current we shall tend to break down the insulation of our cables and any other part of our machinery.

Now I turn on the electric current. By turning the stop-cock I am able to stop it suddenly, and immediately you notice the water escapes through that side tube. If I now open the stop-cock there will naturally be a small leak, through which a very little water trickles; but if I suddenly stop the main flow of water, then the pressure is obviously increased momentarily to a very large extent, producing a spurt of water followed by a steady stream. That is an instructive illustration of the high electromotive force which is generated when we suddenly break the circuit.

I have been introducing to your notice another analogy—that of twisting bodies—as illustrating a line for the transmission of power or for any other purpose. I will show you now what the dielectric strength of this glass rod is. I have here (Fig. 19) a flywheel, suspended by the glass rod, to which I give a somewhat rapid rotation and then a sudden stop [*illustrating*], and you see that the dielectric strength of my conductor was not sufficient to stand the strain, and it has broken down. The strain is produced by the inertia or momentum of the flywheel—that is, by the self-induction of, say, the transformer. I will now put in another rod, and show you how we can prevent these disastrous effects from happening, and that is simply by never breaking the circuit suddenly. By putting in a gradually increasing resistance, the momentum of the flywheel is not able to exert its full impulse instantaneously, and we have a gradual slowing down, and the dielectric is able to stand the strain. To illustrate this, after I have the wheel in rotation, I simply have to grasp it with the fingers and gradually bring it to rest in about a second of time. I have brought it to rest without the slightest undue strain upon the insulation.

That means that, with large and powerful plant, in order to prevent the too sudden breaking of the circuit, you ought to have a switch which will enable you to gradually insert a resistance before you break the circuit altogether. Then the whole danger to the insulation of the system would be done away with entirely.

Now, I have given you these two analogies because my own experience is that the more analogies we have, the greater the help it is to us. I know myself that I have gained the greatest assistance from always putting such problems into a mechanical form; it assists one to foresee results that may happen, and enables one to apply one's mathematics to it. I do not wish it to be understood that by these analogies one can get rid of mathematics. One can illustrate the qualitative results extremely well, but when we come to the quantitative results our true servant is mathematics.

Transformer on Open Circuit.—Lag of Current.

When we are sending an alternating current through a transformer there are two cases that may be considered: one where the power is all being used up, the secondary of the transformer having a full supply of lamps upon it, for instance; and the other where there are no lamps on it, where the secondary is not using up the current at all. In the latter case there is still a sensible current going through the primary of the transformer, and the question is, what becomes of it? Are we using up a large amount of power in the transformer when the lamps are not on? No, we are not using a large amount; we are using a certain amount. The counter electromotive force created by the self-induction of the primary of the transformer mitigates this effect when the transformer is doing no work, as then the electromotive force that is applied to the primary is in opposite phase to the opposing electromotive force of self-induction. The current, which follows very nearly the phase of the impressed electromotive force when the transformer is doing full work, now, no work being done, lags behind the electromotive force, which still further reduces the power consumed in the transformer on open circuit.

What is meant by the lag of the current behind the electromotive force? I have a model here (Fig. 20) which might have been better designed if I had had more time to do it, but it will illustrate what I want to show you. I have a wheel supported on an axis, and I have two weights, one of which in descending causes the wheel to turn in one direction, the other in the opposite direction. What I wish to show you is that when I stop allowing one of these forces to act, the momentum of the wheel will still carry it round in the same direction in which it was going; although an opposing electromotive force is acting, the wheel continues to rotate in the same direction. You will see the analogy which I wish to be conveyed to you. Now I apply the electromotive force, or weight, on the right, and it sets the wheel in motion. Now I stop it, and you notice that the other weight still goes up for a sensible time—that

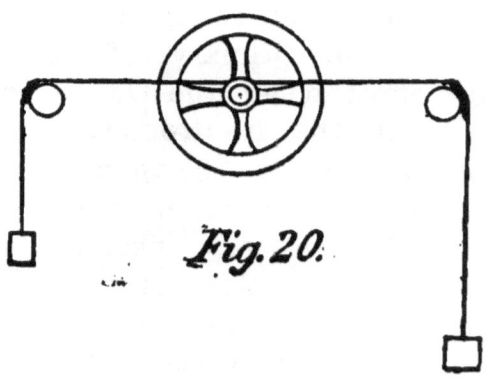

Fig. 20.

is to say, the current lags behind the electromotive force. I might put an alternating electromotive force on first in one direction and then in the other, but the momentum of the flywheel would always be pulling up during a portion of the time against the weight.

Now, I have a very interesting experiment to show you illustrating this lag of the current. It is one which was designed in Prof. Ayrton's laboratory a few years ago by Mr. E. W. Smith, and is a very beautiful experiment. I have here (Fig. 21) two

55-volt lamps connected in series, which will be fed from an alternating source of 100 volts. I shall pass an alternating current through them, and you will see that they light up rather dimly. After that I shall put this high-resistance self-induction coil as a shunt upon one of the lamps, and I shall put on the other lamp a high-resistance non-inductive shunt. The difference between the two conditions of the experiment is this: In the first

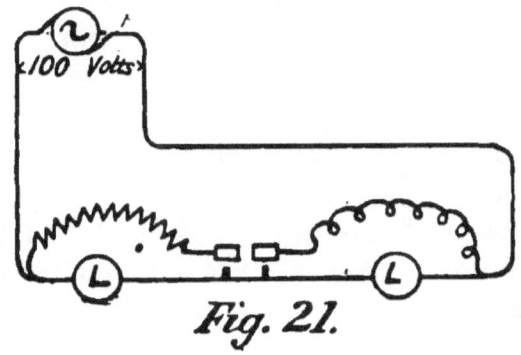

Fig. 21.

place, I pass an alternating current through the two lamps in series, and I get a very low illuminating power. I then give a by-path across the one lamp containing self-induction, and a non-inductive by-path across the other, and you will see that the lamps increase in brightness. [*Experiment.*] The reason is that we have to deal with an additional source of electromotive force besides the main town supply which I am using. This electromotive force is also tending to set up electric currents, and lags behind the phase of the electromotive force which is supplied from the mains; consequently we will have these two lamps lighted up by two currents in different phases, and one will attain its maximum when the other is near its minimum. Of course, we do not see this, because the maxima and minima come a hundred and fifty times a second. So that really by affording by-paths to these two lamps we are able to produce a greater illumination than when the whole current is going directly through them. Of course, we are drawing much larger currents from the mains in this case than we were before, but by introducing these coils we are actually able to increase the brightness of the lamps.

The lag of the current behind the electromotive force is a cause of many curious phenomena, which have been worked out by Prof. Elihu Thomson, of America. Here I have an electromagnet, which I am going to magnetise with an alternating current. It consists of a bundle of iron wire surrounded by a coil. I will show you that the electromagnet produces a repulsion upon conducting bodies brought into its neighbourhood. Of course, the variations of the electromagnetism due to the alternating current will clearly set up induced currents in a ring or any piece of metal brought near to it. But one would have naturally thought that these currents would be strongest just at the moment when there is no magnetism at all in the core, and that therefore they would act sometimes positively and sometimes negatively, that sometimes there would be repulsion and sometimes attraction, and of equal amounts; and that would be the case, if there were no lag in the current. That the reverse is the case, however, is because the current lags behind the electromotive force induced by the lines of force from the electromagnet. The attraction or repulsion, say, on a ring, at any moment, being proportional to the product of the strength of the magnetic field into the strength of the induced current, these former would be of equal amount in an alternation, if there were no self-induction in the ring, and hence no lag, the induced currents being 90 degrees later in phase than the magnetic field. But if there is self-induction in the ring, as of necessity there must be, then the induced currents occur more than 90 degrees later in phase than the magnetic field, and then the products of the strengths of the two at every moment during an alternation show that there is a marked preponderance of repulsive over attractive impulses; the attraction has become comparatively insignificant. The most striking experiment is with an aluminium ring, because of its good conductivity and lightness (Fig. 22). You see, as soon as I put on the current it throws the ring into the air to a considerable height.

I will now show you the same experiment with a copper ring, which is much heavier, but is an extremely good conductor. You will see that it will not rise quite so high as the aluminium ring did. [*Experiment.*]

I will next show you that with worse conducting materials the effect does not take place to the same extent. This is a zinc ring, and you will find that it will not rise to so great a height. [*Experiment.*]

I have here a brass ring, which is even a worse conductor still, and it may possibly be able to topple over the edge of the electromagnet, but that is about all. [*Experiment.*]

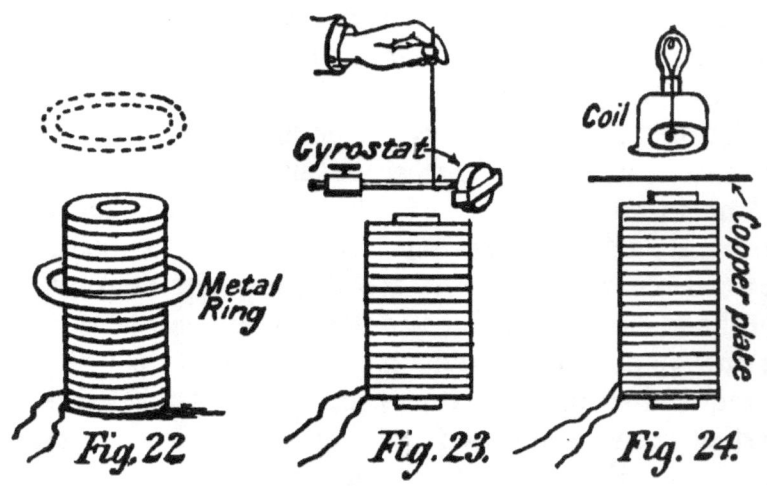

Here is a ring (Fig. 26) held by two threads to a piece of wood, which I can keep floating above the magnet. [*Experiment.*] The ring, of course, gets intensely hot in a very short space of time, owing to the enormous power that is being expended in it.

There are one or two other experiments I can show you. Here (Fig. 23) we have a gyrostat, which consists of an iron wheel, with a thick copper tyre, which becomes magnetised by the induced currents. The axis is supported on a copper framework that surrounds it, and that causes the lag of the current to become operative, and a continuous rotation to be set up in the gyrostat. Here, then, we have an electro-gyrostat driven by this curious force of Prof. Thomson's. He has devised a great number of different methods of illustrating this, from which different results may be obtained. I am able to illuminate this lamp (Fig. 24), but I can screen it off by means of this plate of copper.

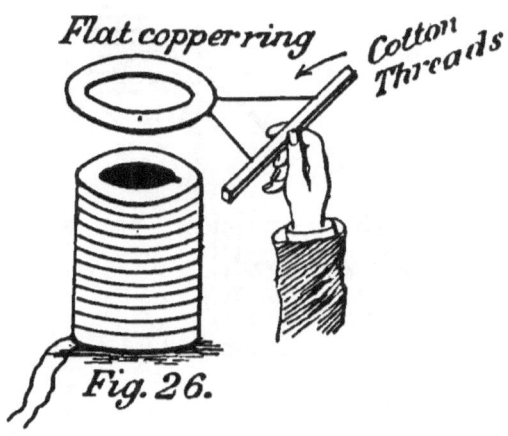

Fig. 26.

It is difficult to pass it between the two coils because the repulsive force is so great. [*Experiment.*] If that had been of a bad conducting material it would not have done it so well, but being a good conductor it screens off the lines of magnetic force, and prevents them going through the coil. The lines of magnetic force, instead of going straight up, are bent round.

Finally, I will show you the effect of trying to

bring a plate of copper upon the pole (Fig. 25). I cannot strike the pole: it always rebounds from

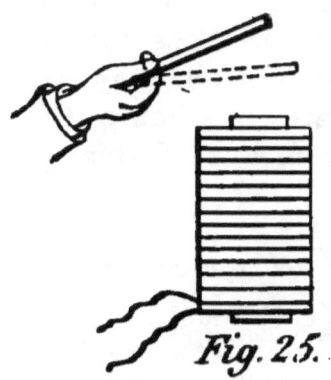

Fig. 25.

it owing to the repulsion caused by the induced currents. [*Experiment.*]

LECTURE III.

Sometimes at the close of a discourse the lecturer finds that he has miscalculated his time, and has to hurry through things more rapidly than he intended, so to prevent this to-day I shall begin by saying a few words which I should have said at the end of the lecture but for the reason just stated. In the course of these lectures I have been able to show you some interesting experiments in different branches of electricity, and also certain curious phenomena connected with them. I wish, in the first place, to say that in all these experiments there is nothing original or new—that they are simply illustrating well-known principles; and, secondly, that I have to express my great indebtedness to all those who have assisted me in the matter—and especially I feel that I owe a great deal to Mr. Campbell Swinton, who has put at my disposal this induction coil and much of the apparatus which you see before you, some of which I shall be using to-day. I wish, also, to express my indebtedness to Dr. Alexander Muirhead for the use of the different illustrations in the matter of the submarine cable, and also to my assistant (Mr. Mitchell), Mr. Stanton, and Mr. Edgar, who have made the work of preparing these lectures quite contrary to what it usually is—a pleasure instead of a labour.

If in any of the illustrations that I have given you of the analogies between electrical phenomena and mechanical phenomena I have introduced anything different from what you have been accustomed to, it is simply the result of a natural interpretation of the mathematical formulæ that represent the phenomena. The same mathematics apply equally well to mechanical phenomena and to electrical phenomena, and it seemed to me a suitable thing on this occasion not to use mathematics to illustrate electrical phenomena, but to use mechanical phenomena to do so.

Practical Utility of Analogies.

Now about these analogies which I have used. I should like to impress upon you before we part that there is more in these analogies than the mere rough resemblance which I have tried to point out to you in the course of these lectures, and those who are more deeply interested in the applications of electricity will easily see that there is room for a great deal of thought in carrying out these analogies, and in making clearer to their minds the processes which are going on in certain electrical phenomena.

In the table which I showed you last lecture I gave you an example of the methods of measuring the actual units when we are using the twisting of any rod or other mass to represent electric currents, and I wish to say that if in the case of any electrical problem occurring to an engineer —such, for example, as the propagation of electricity along a cable from Niagara Falls to Buffalo— he wishes to see whether troubles will arise in the electrical phenomena, if he will make an imitation cable on the lines of assuming those units which are in the table, he will be able to imitate the action of what will go on in the real cable, and he will be able to see whether any peculiar mechanical effects are produced, and to trace their analogy with the electrical effects, and so be able to see what kind of troubles he is likely to meet with. And this I consider to be a matter of very great importance, but it was with the object more of throwing out suggestions than of working out the analogies in a complete form that I introduced them in the present course of lectures. The working of them out in a complete form would not have been a suitable subject for these lectures, but I hope I may have time on another occasion to work some of them out more fully, and so to render some assistance to those who are engaged in such work.

Further Alternating-Current Phenomena.

In the first lecture I was dealing chiefly with capacity and the influence which it has upon alternating or interrupted currents, and the most interesting illustrations of the effects of capacity were naturally found in the submarine cable. Last lecture I was dealing chiefly with self-induction, and I showed you that there were certain troubles which were likely to arise in cases in which we used self-induction, and I explained to you that this would be so with both kinds of currents—whether we were using continuous currents or alternating currents; that in either case, if you have much self-induction in the circuit, by breaking the circuit suddenly you are able to get far larger electromotive forces than those you are impressing upon the system, and that these may be destructive to the insulation of the system. The only difference is that, in the case of alternating currents, the self-inductions that we have—for example, in the transformers—are enormously greater than those existing when using continuous currents; and, therefore, the importance of this question becomes clearly greater in the practical application of alternating currents, and especially in the practical application of them on a gigantic scale, as in the case of the utilisation of the Niagara Falls, where the transformers are of enormous magnitude, and the self-induction very great indeed. In such a case as that, it would be, I might almost say, criminal to suddenly break the circuit, the danger would be so great.

Now, to-day, passing from these two subjects, I am going to deal with three other peculiarities which arise in the transmission of electric currents, especially alternating or interrupted currents—skin resistance, resonance, and oscillations. All of these are interesting phenomena from a purely scientific

ELECTRIC CURRENTS. 73

point of view, and are of far more importance from an engineering point of view. I mentioned in the course of the last lecture that a great many of the difficulties presenting themselves in discussing an engineering problem can be got over by reducing the frequency of the alternations. When I first reported to the people who are utilising the Niagara Falls, in the year 1890, now some five years ago, as to the proper means of utilising them, I urged the adoption of the system we are now introducing; and on noticing those various troubles that might arise, I gave the best solutions of them that seemed available. Since that date the enormously valuable suggestion occurred to me of lowering the frequency of the alternations, and that, together with the other precautions that have been introduced, has taken away all the danger of any trouble that we might otherwise have expected from the use of alternating currents with such a system.

Induction with very High Frequency and Electromotive Force.

When we come to very high frequencies, effects are greatly increased; and by means of Mr. Campbell Swinton's apparatus which I have been employing I am able to get very high frequencies indeed. By using an oscillatory discharge (about which I shall speak more fully later on) which oscillates at very rapid frequency, and also by using high electromotive force in this

F

discharge, I am able to get effects which could not be got by an ordinary alternating current. At the same time, by observing the effects with these very high frequencies, we are able to grasp what is going on at the lower frequencies, though not to such a serious extent.

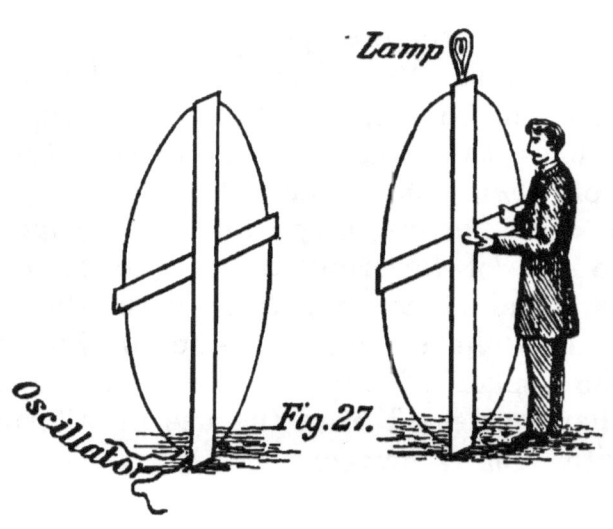

Fig. 27.

I will now show you the effects of mutual induction between two circuits, which can be obtained by means of very high frequencies and by very high electrical pressure—great voltage. I am going to employ the two copper wire hoops (Fig. 27), measuring about 6½ or 7 feet in diameter, which are standing in front of me, and I am going to pass through one of them an oscillatory discharge of very high frequency indeed and of very high electrical pressure. Now, everyone who is accustomed to seeing cases of mutual induction will be perfectly well aware that if we are simply passing an alternating current through one of these wires, it would be impossible to induce much current in the other wire. I wish to show what a considerable amount of power can be developed when the electromotive force is sufficiently high and the frequency sufficiently great. And that is what obtains when we use the oscillatory discharge of Mr. Campbell Swinton. I take away from the apparatus one of the hoops, to which I have attached an incandescent lamp at the top. I will immediately cause a high-frequency oscillatory discharge to pass round the other, and I wish to show you that even at a considerable distance the current which passes through the wire which I am holding is sufficient to light the incandescent lamp at the top. [*Experiment.*] At the present moment the illumination is not visible. I approach the hoop nearer to the other. At this distance it is very dull indeed, but preceptible. I now approach to within a couple of feet, and you see it rises to its full brilliancy. That is simply due to the induced current which is going round this wire of a diameter of 7 feet. I can twist it round in different directions, almost at right angles to the fixed one, and still get a certain amount of illumination, but when it is completely at right angles it goes out. When I turn further round and the two wires are once more parallel, the brightness is as great as it was before.

Lighting Lamp by Current passing through a Person's Body.

Now, I intend to take some of the same apparatus which I have been using here to show you another very remarkable effect of these very high-voltage experiments. When we use very high electric pressures we require a much smaller current to produce the energy, which lights up an electric lamp, for instance; and if we send very high electric pressures through a lamp with intervals of rest, when the current is not passing, we are able to produce a very large current for a small fraction of time, after an interval another very large current, and on the whole the total quantity of electricity which has passed through to produce the lighting of the lamp will not be so great as the total quantity which would have passed through if we had had an ordinary continuous or alternating current passing through the lamp. The effect which takes place on the lamp is measured by the square of the current. Let us take two cases—first, of a lamp with the current continuously going through it: we have a certain amount of electricity passing through the lamp; in the second case, if we pass it for one-tenth of a second, each second of time, and pass ten times as much current through it, we would get one hundred times as much effect during that tenth of a second: therefore you would get ten times as much effect on the whole. It follows, then, that when we increase the maximum current during a short period we are increasing the energy which is supplied to the lamp during that period far more than we are increasing the current. It appears, therefore, that by acting in this way we are able to pass a far smaller sum total of current through the lamp, and yet to get the same amount of energy consumed in the lamp. This is probably the reason why an electric lamp can be lighted by a current passing through a person's body, as is possible with such apparatus as is before you. The amount of current which I am able to stand without injury from hand to hand is certainly less than 1-100th of an ampere; and if I can show that a lamp can be lighted through my body, I know that there is not in the energy more than 1-100th of an ampere passing through the lamp, and yet if I bring the lamp up to incandescence—a lamp that ought to require one-fifth of an ampere—I must assume that the energy is the same, but that the total quantity of electricity is not the same; and that is what I wish to show you is the case with this apparatus. Now, this is an experiment which everybody does not like to try, so I will ask Mr. Campbell Swinton, who has arranged it, if he will assist me in taking the current through the body. There is really not the slightest trouble, or danger, or inconvenience in doing it. I showed you in the first lecture some experiments of the same sort. There it was perfectly obvious that the electricity which was lighting the lamp was not passing through the body, but in this case it is perfectly obvious that the electricity is passing through the body (Fig. 28). [*Experiment.*] The electricity coming from that coil is certainly going through both of our bodies, and we can lower it to any amount we please, or raise it up to its full incandescence, by cutting out more or less of the impedance coil, shown across the centre of the diagram.

Whether the explanation which I have given you, and which Mr. Campbell Swinton supports, is the

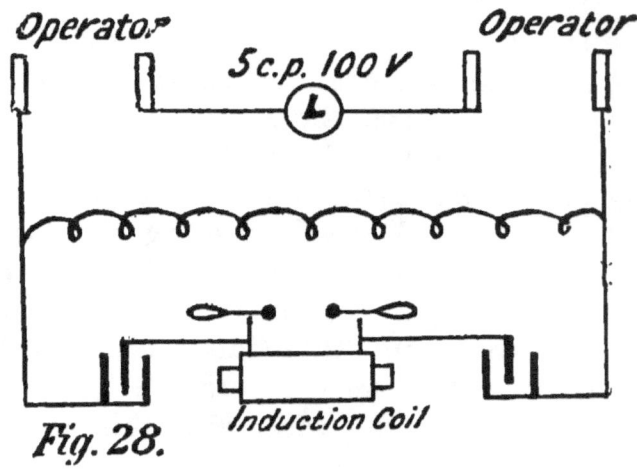

Fig. 28.

complete explanation or not I cannot tell you. It seems to go a long way towards explaining it. Passing the current through the body is a very different thing to receiving it, using yourself as a Leyden jar, as I did in the first lecture. There, there is no sensation whatever; you do not feel the current at all. In this case both Mr. Campbell Swinton and I felt the current passing through our bodies—and it was nearly as much as one cared to take comfortably, and one feels a certain amount of warmth, but I do not think that that is the true calorific value which is imparted owing to the energy used up in one's body; but it certainly does warm one.

Skin Resistance.

I now wish to introduce to your notice a subject which has attracted considerable attention during the last eight or nine years of the use of alternating or interrupted currents. I speak of what has gone by the name of "skin resistance." The fact has certainly been proved—and it could not be otherwise, from the mathematical treatment of the question—that, when we pass a high-frequency alternating current through a conductor, the flow of current is not uniform throughout the section of the conductor. It is, partially at least, confined to the exterior layers of the conductor. At the centre there is not so much current passing as there is through an equal section of the outside, and therefore the resistance of the conductor is composed more of the resistance of the external parts than of the central part. If we make the frequency high enough, we can cause the current to cease flowing through the central part entirely. In that case, the material which is offering resistance is simply a tubular conductor and not a solid conductor. That is the meaning in which the term "skin resistance" must be read.

It was in the presidential address to the Institution of Electrical Engineers in, I believe, the year 1886, that Prof. Hughes first introduced this seriously to the notice of electricians. I do not mean to say that some of the effects had not been observed before, and I do not say that it was not known that there might be difference of flow in different parts of the section of the conductor, but that was the first date when this resistance was proved to consist of a pure, definite resistance, and was measured in ohms as a resistance by Prof. Hughes. Up to that time people had been perfectly well aware that, in using the Wheatstone bridge, if you kept the key of the galvanometer (a deflection of which indicates divergence from the resistance) down, and interrupted the circuit from the battery, you could not get a true reading. Generally, this was ascribed to self-induction, which at the time of breaking the circuit to the electric battery made itself apparent, but which to a far greater extent, except when you were using coils, was often due to what Prof. Hughes showed to be really an increase of the resistance; and that increased resistance is due to the electric current being at the outside instead of being uniformly distributed through the section of the conductor which is carrying the current. Well, the explanation of this is not difficult to seek. You remember in the last lecture I had two wires of about 7 feet long parallel to each other. I passed an electric current through one, and at the moment of making that circuit the other, which was completed through the galvanometer, showed an induced electric current, and at the moment of breaking the primary current the circuit connected with the galvanometer again showed an induced current, but in the opposite direction to that previously exhibited. Thus, with an alternating current, at the moment when a change in the direction of the electric current takes place, there is a reverse effect due to mutual induction. Now, suppose that this rod on the table (which is merely for illustration) consisted of a bundle of wires, and I insulated and separated out the central wire, and also insulated and separated out one of the external wires, and then sent through the whole bundle an electric current, there would be an induction current in both of those which I have selected and insulated ; there would be in those an opposing electromotive force, a force tending in the opposite direction to that in which the primary current was going. But if you will notice the position of all the wires in the bundle relatively to one of them on the outside, and also notice the position of them relatively to the central wire, you will see that the central one has far more wires within a short radius of it than the external one has. Therefore the opposing force on the central wire is much greater than the opposing force on the external one, so that, when I suddenly pass an alternating current through a bundle of wires, at each change in the direction of the current there is an opposing force in both these wires, but the opposing force in the central wire will be much greater than that in an outside wire. That is to say, there is an opposing force in each of the wires of the bundle, but a greater opposing force in the central ones, and therefore the current will not flow so powerfully there as it does at the outside. If, therefore, we have the whole bundle grouped into a solid conductor, the central line has always an opposing electromotive force which is greater than the opposing electromotive force at any other portion of the section. Therefore the current is mainly passing through the external part of the conductor, and that causes an increase of resistance, which was actually measured by Prof. Hughes by means of the ingenious apparatus he produced on the occasion to which I have referred.

Here is a brass tube, with which, however, it would take some time to show you the effect, because it needs a very sensitive galvanometer, and it takes some time to bring it to rest. Prof. Hughes showed that the additional resistance produced by an iron conductor was much higher than a conductor made of any other material. Therefore, I will proceed by using an iron tubular conductor (Fig. 29) with a wire supported through its centre, and another insulated wire lying on its surface, and I wish to show you the different effects of induction on these two. I will begin with the outside wire, which is subjected to the induction at varying distances up to the diameter of the tube. There will only be a deflection in this case of two or three divisions of the scale on the screen, so you will please watch carefully and make the measurements for yourselves. [*Experiment.*] There, you see, it goes nearly up to the second division. I passed the current through this mass of iron, and the induced current on the outside wire was sufficient to deflect it through about one and a half divisions of the scale. Now change the connections, and you will see the great difference there is. [*Experiment.*] This time you see that the deflection passed along nearly the whole length of the scale. That is the deflection

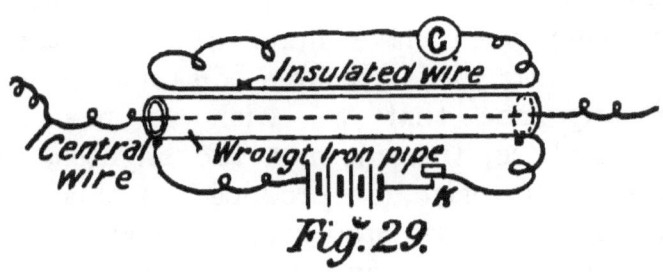

Fig. 29.

obtained with the wire which passes through the centre of the tube, showing the enormous difference of the electromotive force in the two parts. This enormous effect in the case of the iron is due to the fact that the iron becomes magnetised by the passage of the electric current through it— it becomes circularly magnetised. Lines of force cut the wire which is suspended in the middle of the tube, and in cutting it they exert a force greater than the electromotive force that there would be if there had been no iron—for example, if the iron had been replaced by copper, through which you cannot get so great an amount of magnetism so easily. This illustrates the general principle of skin resistance. I may say that in actual practice an engineer is seldom troubled with this except when he is coming to very large works where the cables are very thick indeed. With all the cables that we have been accustomed to in industrial work in this country there should be no trouble whatever ; and it is only when you come to cables of one or two inches in diameter that the trouble would come to be serious. But when once you lower the frequency as I have done on the other side of the Atlantic, this trouble disappears entirely. Up to any size that we wish to use for our conductors there, there is not the slightest trouble from the extra resistance produced in this way.

I can bring before you some other illustrations of the same sort to show you that the skin resistance may become very high indeed, and may even almost prevent an electric current from passing through a good solid conductor. This is really an explanation of the experiment which was made by Faraday so long ago, in which he had a wire going round the room and separated by a few inches where the two ends came together, into which an electrical discharge from a powerful Leyden jar was sent at the two ends, when the discharge preferred to cross the air-space rather than to travel round the copper conductor, although that was an excellent conductor. That was owing

ELECTRIC CURRENTS. 85

to the suddenness of the discharge, and the experiments I am going to show you are illustrative of the same fact. To show you it in a more complete manner, perhaps, I will take a copper wire arch (Fig. 30) about 6 feet high—this is one of Mr. Campbell Swinton's arrangements for which I am indebted to him—and I will send into it an oscillatory discharge. I have the means of producing very rapid oscillations in the apparatus before you.

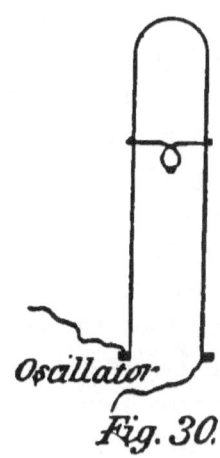

Fig. 30.

The terminals will be connected with the two ends of the arch at the base, and there is an incandescent lamp stretched across the two vertical pillars which form the sides of the arch. Any person looking at this in relation to the ordinary method of using currents would say that, when we applied the current, it would all flow through the thick arch and none would flow across this very high-resistance lamp; but, as a matter of fact, the resistance opposed to these very high oscillations becomes so great that it is very difficult to get the current to go up a short length of this wire, and still more difficult to get it to go right round; and instead of going right round by that thick conductor, which would offer a certain amount of skin resistance, a great portion of the energy prefers to go across that thin filament, where the skin resistance is not so predominant a feature. [*Experiment.*] At this height, you see, comparatively little current is able to go through the lamp. When we bring the lamp lower down, nearer to the terminals, we get more and more of the current able to pass through the conductors up as far as the lamp, and less and less able to go round the rest of the copper arch, and when we bring it down low enough we are able to get the lamp up to its full incandescence.

Now that I have explained to you the nature of skin resistance, and I have explained how it was produced owing to there being a solid conductor, in order to demonstrate this to you better I will now take a strip of copper instead of a piece of solid circular section. This strip of copper is of exactly the same weight per unit length, it is of the same section, as the wire, and it is of the same conductivity. And what I wish to show you here is that the induction of one part on another is not such a serious matter in the case of the strip, where the distances from the centre of the various parts in a section are greater than in the case of the wire, where they are all very close to the central part of the section. We will now put on the current. There, you see that the current is able to pass through the copper strip much more easily at the top than it was before. The light is very dim, but as I gradually lower the lamp, gradually there is more light. The lamp, however, is not yet at its full incandescence, and it is only when I bring it down near to the base of the arch that I am able to get sufficient current through the lamp to bring it to its full brilliancy. The difference between the copper strip and the solid copper rod is most marked.

Next I will show you the effect with an iron rod, and in this case I have chosen an iron rod of such a section that it has exactly the same conductivity as either the copper rod which I first used or the copper strip which I used in the second place. And here the opposition to the passage of a current is very much greater than in either of the cases which we had before, as will be obvious as soon as I put the thing into action. [*Experiment.*] Gradually, as we lower it, it is coming towards its full brilliancy; but, you see, when the lamp reaches the same level at which in the first case it arrived at its full brilliancy, it has not arrived at its full brilliancy in this case, simply for the reason that the iron between the lamp and the terminals offers such an amount of opposition that there is very little current able to pass through the lamp even at that height, and we have to lower it very low indeed before it arrives at its full brilliancy. These three experiments illustrate in an extremely good way the general character of the phenomena of skin resistance, which not only from a purely scientific point of view are of great interest, but also require careful consideration in dealing with engineering problems.

Electrical Resonance.

Passing now from skin resistance, I wish to talk a little—I cannot very well show you electrical experiments, although I shall show you experiments illustrative of electrical oscillations—about the phenomena of electrical resonance, such as the rise in pressure noted some time ago on the conductors running from Deptford to London, owing to the considerable capacity of the mains used. You will see the distinction I draw between effects of resonance and effects of oscillations. Let me introduce the subject by saying a few words about ordinary resonance. I have here (Fig. 33) a spiral spring, suspended from the ceiling, to which I can apply a force at regular intervals. And I will show you that when I apply a very small force to the spring in a vertical direction at regular intervals it accumulates the energy, and eventually gives rise itself to a much greater force than that which I had been imparting to it at each individual motion. And also you will see that the velocity of motion after I have repeatedly given an impulse to the spring becomes very much greater than it would be with the same force applied at a single effort. The reason of this is that the spring is capable of vibrating in a perfectly definite period. The period of vibration depends upon two things—it depends, first, upon the elasticity of the spring, which tends to draw it back when once it has been displaced; and, secondly, it depends upon the mass or inertia of the spring, which resists

Fig. 33.

the movement. These are the only two things that come into play in this spring. There is no friction. It is not in a vessel of oil or anything of that nature which would offer resistance. The electrical analogy is simply a circuit in which there is capacity—that is, the elasticity part of it—and self-induction—that is, the inertia part of it. Just as with the mechanical spring, when we know the

G

mass and elasticity of it, we can calculate the period, so, substituting for mass and elasticity the electrical units which I have given in the table (page 42), you can calculate what the period of an electrical vibration is also. [*Experiment.*] I exert a very slight force upon it at each stroke, and, you see, before I have taken more than two or three strokes its motion is increased to a very considerable extent. The vibration of the spring now in its own period is very violent indeed—that

Fig. 31.

is to say, the current that is flowing through that electrical system is very much greater than that which is due to the electromotive force. That is the real interpretation of the increased velocity of the spring.

I will now give you another illustration of electrical resonance. I take for that purpose a jar (Fig. 31) containing a little water, and also a certain amount of air. The air is capable of vibrating, and if the length of the jar is fixed it is capable of vibrating in a perfectly definite period. When I cause a tuning-fork to vibrate, and hold it over the mouth of the jar, it sets the air at the upper part of the jar vibrating ; this vibration communicates itself to the air in the lower part, and it is then reflected from the bottom just in exactly the same way as that spring was from the ceiling. The air at the top of the jar represents the free end, and the water, being fixed, corresponds to the ceiling. I will now show you that the air in the jar will resound to this tuning-fork, and greatly increase its sound. The period that it will resound in depends upon the length of the column of air. [*Experiment.*]

I used then a tuning-fork of low note. Here I have one of a higher note, and you will see that this column of air will no longer resound to it, and we must find a column of air of different length which will do so. I will cause the tuning-fork to vibrate, and then get Mr. Mitchell gradually to fill up the jar with water. As the water is poured in you will find that the sound gradually increases, until when we reach a certain length the resonance is complete. [*Experiment.*] Knowing the elasticity of the air, and knowing the weight of the air, we can calculate the period just the same as we can in the case of the spring, and so we can in any electrical system where we know the farads and the henrys. As in the cases of the spring and the air, the period depends only on the elasticity and inertia, so in the electrical phenomenon it depends on the capacity and self-induction, and not upon the resistance. The resistance may tend to dim the resonance, but does not alter its period.

As an illustration of ordinary resonance, I will show you an apparatus (Fig. 32) which I got out some time ago for measuring the amount of firedamp in a coal mine. It consists of a tube, whose length I can vary by means of a wheel, and a tuning-fork which resounds to it. [*Experiment.*]

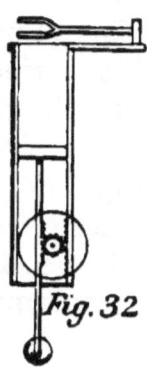

Fig. 32

There is no resonance now; there is now; there is none now. I pass through a maximum of resonance. I adjust the length of the tube by turning the wheel till I get the maximum, and I read by the index that there is firedamp, or there is not, as the case may be. The elasticity of the firedamp is the same as the elasticity of the air, but the mass that is to be moved is not so great, and therefore it will move more rapidly than a column of air of the same length.

Electrical Oscillations.

Leaving the question of electrical resonance now,

ELECTRIC CURRENTS 93

the last feature of interrupted and alternating currents which I wish to show you is oscillations. In order to illustrate the phenomena, I have prepared a model to carry out our analogies. The oscillators which I am going to use consist of a capacity and a self-induction which are so chosen as to give a very rapid period of oscillation, so that effects are produced by an alternating current with

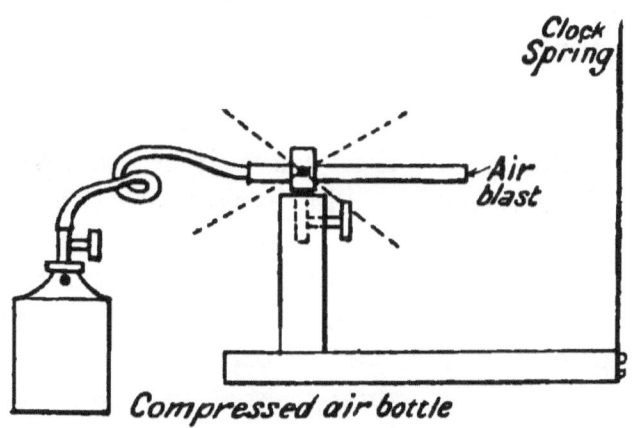

Fig. 34.

these oscillators which could not be produced by the alternating currents alone. I will now show you by means of a simple model what the character of these is. I have a spring which you see here (Fig. 34). It has a natural period of oscillation. I have applied a strip of paper to the piece of clock-spring, so that I can blow against it and produce a disturbance. The velocity of the motion represents the amperes of the current; and what

I want to show you is that I can get far more powerful currents from this by a broken current than I can by using an undulatory current. I have arranged it so that in the first place I move a jet of air up and down so as to be undulatory in its character. [*Experiment.*] There, now notice the amount of the motion and the amount of the velocity. You notice the spring moves backwards and forwards as the air jet is moved, and the velocity is comparatively slow.

Now, I am going to use the same jet of air, but instead of passing it so as to be continuously pressing upon it, but with more or less effect, I am going to pass it across the spring so as to be acting upon the spring only for a minute space of time, and then leave it to its own oscillations. [*Experiment.*] There, you see the velocity is one hundred times as great as it was before, although the amount of pressure which I am applying summed up over a second is not nearly so great, and yet I get enormously more velocity than I did before, when I was applying an undulatory current. In the electrical analogy, if I have an oscillating system I am able to produce a far greater current by allowing it freely to oscillate than if I apply a gradually increasing and diminishing current to the system.

As an illustration of this I will show you an experiment with the oscillatory discharge. These oscillatory discharges of high voltages have been used so frequently of late years that I suppose most people have seen them. At the same time

ELECTRIC CURRENTS. 95

they are very instructive experiments, and it will take but a few moments to arrange them. I just wish to show you the lighting up of a tube (Fig. 35). The experiment was first shown in this room extremely well by Mr. Nikola Tesla, of America, some few years ago, and it has justly gone by his name ever since. The oscillations are transmitted through my body and are able to light up the tube. [*Experiment.*] Here, you see the tube lights up through my body, and the oscillations are shown by the intervals of light.

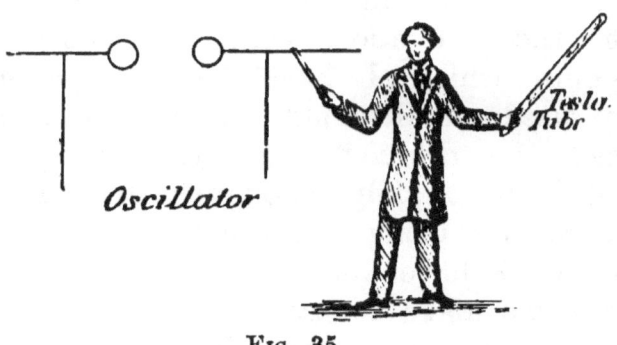

Fig. 35.

It lights up only when the discharge has taken place, and you are thus enabled to see the nature of the oscillatory discharge.

Experiment by Lord Armstrong.

I have now come to the end of the experiments which illustrate my subjects, and I feel sure that those of you who have listened to these lectures, and have seen the admirable arrangements which Mr. Campbell Swinton has made, will not be disappointed if I show you another experiment, using the same apparatus, but not connected with the subject of the lecture. The experiment is a very remarkable one, and was originally due to Lord Armstrong. He first found it out when using a hydro-electric machine in which he got great currents at high pressures by means of a jet of steam. In the arrangements which I am showing you I get the high electromotive force, or rather Mr. Campbell Swinton has obtained it, by using an induction coil and using a continuous current through the induction coil. In most of the experiments which I have been showing before we have used an alternating current, because if we had used a continuous one we could not have interrupted it rapidly enough. In the present instance we are using a continuous current. From experiments I have shown you before, you know that when we make contact with the primary there is an induced current in the secondary, and a reverse one at break. We are able to almost eliminate the current in one direction, and to thus secure the secondary current almost entirely in one direction; and this is emphasised by having a spark gap which the current has to leap over, and which is wide enough to prevent the spark passing at the make, but not too wide to prevent it passing at the break. This is really an experiment on continuous currents, and not on either alternating or interrupted currents.

In this experiment the apparatus (Fig. 36) is not shown on the screen inverted; it is shown in the same position as it is on the table in front of the lamp. Here, in a square glass trough, I have a glass bulb, open at the top and open at the bottom, with a piece of cotton wick coming through the narrow neck-like opening at the bottom of the bulb. The wick is at present coiled up in the bulb, and a short end is projecting into the liquid below. The high-pressure electric current is taken to

Fig. 36.

the water outside by one terminal, the other terminal being in the water inside the bulb. According as I pass the current in the one direction or the other, so the cotton wick will be seen to move out of the bulb, or to be retracted again into the bulb; and the most remarkable feature about this is that at the same time that the cotton wick is moving in one direction the water is moving in the other, and by the projection on the screen you will actually be able to see the direction in which the water is moving. In order to carry out this experiment, the purest distilled water is necessary, and provided everything goes well we ought to see the motions easily.

I may say that Lord Armstrong has shown this experiment to a few of his friends, and Mr. Campbell Swinton has shown it to me and others, but very few persons have seen it. I can only say, in explanation, that it is probably due to the same phenomena as electrical endosmose. If you have a vessel of liquid with a septum of porous material, the terminals of a battery being immersed in the two compartments, there is an action which tends to draw the liquid through the septum, it being probably a mutual action between the porous material and the water. Probably the same effect is exhibited in this experiment; there is a mutual action between this wick and the water outside, and the mutual action causes the one to be drawn downwards and the other to be pushed upwards. [*Experiment.*] Now you will notice that the water is rising in the bulb, and the wick is coming out at the bottom. We will reverse the current, and as soon as we do that you will see that the action takes place in the opposite direction, and the wick is going upwards while the water is going out at the bottom.

INDEX.

A.

Alternating Currents, 22, 50, 57, 60, 71, 75, 81, 93
Frequency 23, 56, 73-75, 84
Phenomena 72
Aluminium Ring Experiment 65
Ampere, Mechanical Analogy of 42, 93
Analogies—Ampere 42, 93
Atlantic Cable 11, 25, 40, 41
Coulomb 42
Electrical Oscillations 93
Electric Circuit 19
Henry 42
Incandescent Lamps, etc. .. 12
Lag 61-65, 67
Mechanical 25-28
Mechanical, Deductions from 38, 39
Mutual Induction 53-55
Pressure 12
Self-Induction 55-59
Submarine Cables 28-33
Utility of 9 11, 13, 70, 71
Volt 42
Armstrong's, Lord, High-Pressure Experiment 96

B.

Brass Ring Experiment 66
Breaking Circuit 59

C.

Cables—Current through 11
Submarine 11
Submarine, Analogy of 28-33
Submarine, Working Torsion Model of 35, 37
Capacity 19, 21-25, 28
Chain, Rotating 16, 18
Clerk Maxwell's Work ... 39, 40, 53
Condensers 22
Copper Ring Experiment 65
Coulomb, Mechanical Analogy of 42
Counter Electromotive Force, 61, 82
Curb Sender 33, 34
Currents—Alternating, 22, 50, 57, 60, 71, 75, 81, 93
Broken 94
Continuous 20, 50, 51, 96-98
Induced 66-68, 81
Undulatory 94

D.

Discharge Experiments, Faraday's 19
Discharge, Oscillatory, 73-75, 94, 95
Disc, Rotating, Experiment 18

E.

Electrical Resonance 88-92
Electricity, Propagation of 9, 11, 71
Electromagnetics, Primary
  Discovery of 43
Electromagnetism, Variations
  of 64
Electromotive Force, Counter
  or Opposing 61, 82
Energy, Transmitting 49
Experiments — Armstrong's,
  Lord 98
Capacity 21, 27
Curbing 32
Current, Retardation of ... 30
Current Lag, etc., 62, 63, 65 67
Electrical Oscillations . 94, 95
Electrical Resonance 90, 92
Faraday's 43, 84
  with Gramme Ring 47
  with Gyrostats 15, 66
High-Voltage 75-79
Oersted's 43
Skin Resistance 83-88
Smith's, Mr. E. W 62
Syphon-Recorder 33

F.

Faraday, Work of, etc., 22, 43,
  48, 84
Farad, Mechanical Analogy of 42
Force, Electromotive, Illus-
  tration of 59
Frequency 23, 56, 73-75, 84

G.

Gramme Ring Experiment ... 47
Gyrostat, Experiments with, 14,
  15, 66

H.

Henry, Mechanical Analogy of 42
Hughes, Prof. 34, 80, 82
Hydro-Electric Machine 96

I.

Incandescent Lamp, Analogue
  of 12
Experiments with, 23, 47 63,
  75-78
Induction.. .... 43, 44
Coil 51, 52
Mutual 50-55, 74, 75, 81
Self 18, 44-48, 56, 57, 65 72
Inertia 44 45
Electrical 18

K.

Kelvin's, Lord, Ether Theory 55
  Views on Power Transmission at Niagara Falls..... 50

L.

Lag, Current 61-65, 67
Lamps, Incandescent, 12 23, 47,
  63, 75-78
Light, Arc, Experiment with 52

M.

Mechanical Analogies, Use-
  fulness of 13
Muirhead's, Dr., Cables etc., 10,
  30, 69
Mutual Induction 50-55

INDEX.

N.

Niagara Falls... 10, 49, 50, 56, 57, 71-73

O.

Obvious Mechanical Facts ...	13
Oersted's Discovery	43
Oscillations, Electrical.........	92

P.

Potential, Analogy of Unit of 42
Pumps, Centrifugal, Analogue of Pressure by... 12

R

Resistance, Experiment ... 82, 83
Skin 79, 86-88
Resonance, Electrical......... 88-92

S.

Self-Induction, 18, 44-48, 56, 57, 72
Shunts 63
Smith's, Mr. E. W., Experiment 62
Spring, Period of Vibration of 88, 89
Swinton's, Mr. Campbell, Apparatus, 10, 69, 73, 75, 77, 78, 85, 96

Syphon-Recorder 33

T.

Tesla's, Mr. Nikola, Experiment	95
Thomson's, Prof., Force ...	64, 67
Transformers......... 48, 49, 60,	61
Transmission	72
Transmitter, Cable	33

U.

Units, Electrical Analogies of, Compared with Mechanical, 41-44
Measuring..................... ... 70

V.

Volts, Mechanical Analogy of 42

W.

Water Analogue of Electrical Pressure	12
Wheatstone Bridge	80

Z.

Zinc Ring Experiment 66

BIGGS & CO.'S BOOK-LIST.

ELECTRICAL ENGINEERS' SERIES.

First Principles of Electrical Engineering.
Second Edition. Crown 8vo. Price 2s. 6d. By C. H. W. BIGGS, Editor of *The Electrical Engineer* and *The Contract Journal.*

First Principles of Mechanical Engineering.
By JOHN IMRAY, with additions by C. H. W. BIGGS. Illustrated. Price 3s. 6d.

First Principles of Building : Being a Practical Handbook for Technical Students. By ALEX. BLACK, C.E. Illustrated. Price 3s. 6d.

First Principles of the Locomotive. By MICHAEL REYNOLDS. Illustrated. Price 2s. 6d.

Portative Electricity. By J. T. NIBLETT, author of "Secondary Batteries." Illustrated. Price 2s. 6d.

Popular Electric Lighting. By CAPTAIN E. IRONSIDE BAX. Illustrated. Price 2s.

Practical Electrical Engineering. Being a complete treatise on the Construction and Management of Electrical Apparatus as used in Electric Lighting and the Electric Transmission of Power. 2 Vols. Imp. Quarto. By Various Authors. With many Hundreds of Illustrations. Price £2. 2s.

Dynamos, Alternators, and Transformers. By GISBERT KAPP, M.Inst.C.E., M.Inst.E.E. Fully Illustrated. Price 10s. 6d. The book gives an exposition of the general principles underlying the construction of Dynamo-Electric Apparatus without the use of high mathematics and complicated methods of investigation, thus enabling the average engineering student and the average electrical engineer, even without previous knowledge, to easily follow the subject.

Electric Traction. By A. RECKENZAUN, M.I.E.E. Illustrated. Price 10s. 6d. This is the first serious attempt to consolidate and systematise the information of an important subject. Mr. Reckenzaun's experience is of the longest and widest, and this book deals not only with the scientific and practical problems met with in traction work, but enters somewhat into the financial aspect of the question.

Electrical Distribution: Its Theory and Practice
Part I.: By MARTIN HAMILTON KILGOUR. Part II.: By H. SWAN and C. H. W. BIGGS. Illustrated. Price 10s. 6d.

Electric Light and Power: Giving the Results of Practical Experience in Central-Station Work. By ARTHUR F. GUY, A.M.I.C.E. Illustrated. Price 5s.

Alternate - Current Transformer Design. By R. W. WEEKES, Whit.Sch., A.M.I.C.E. Crown 8vo. Illustrated. Price 2s. This book is one of a new series intended to show engineers and manufacturers the exact method of using our acquired knowledge in the design and construction of apparatus.

Town Councillors' Handbook to Electric Lighting.
By N. SCOTT RUSSELL, M.Inst.C.E. Crown 8vo. Illustrated. Cloth, 1s.

Physical Units. By MAGNUS MACLEAN, M.A., D Sc., F.R.S.E. Illustrated. Price 2s. 6d. This little book discusses the present state of the subject under the following headings: Fundamental Units—Geometrical and Kinematical Units—Dynamical Units—Electrostatic System of Units—Magnetic Units—Electromagnetic System of Units—Practical Electrical Units.

Secondary Batteries. By J. T. NIBLETT. Illustrated. The Second Edition of this Work will shortly be issued, the price of which will be 5s.

Theory and Practice of Electro - Deposition:
Including every known mode of Depositing Metals, Preparing Metals for Immersion, Taking Moulds, and Rendering them Conducting. By DR. G. GORE, F.R.S. Crown 8vo. Illustrated. Price 1s. 6d.

Alternating and Interrupted Electric Currents.
By PROF. GEORGE FORBES, M.A., F.R.S., M.I.C.E. Illustrated. Price 2s. 6d.

Economics of Iron and Steel. By H. J. SKELTON. Illustrated. Price 5s.

IN THE PRESS.

First Principles of Electricity and Magnetism
By C. H. W. BIGGS, M.I.E.E., Editor of *The Electrical Engineer*. Illustrated. Price 2s. 6d. This book has been prepared more especially to assist students for the elementary examinations for the City and Guilds Institute.

Central - Station Management. By J. HESKETH, M.I.E.E., etc., Borough Electrical Engineer, Blackpool.

MUNICIPAL ENGINEERS' SERIES.

Refuse Destructors, with Results up to Present Time. Second and Revised Edition. A Handbook for Municipal Officers, Town Councillors, and others interested in Town Sanitation. By CHARLES JONES, M.Inst.C.E., Hon. Sec. and Past-President of the Incorporated Association of Municipal and County Engineers; Surveyor to the Ealing Local Board. With a Paper on "The Utilisation of Town Refuse for Power Production," by THOMAS TOMLINSON, B.E., A.M.I.C.E. With numerous Diagrams. Price 5s.

The Contractors' Price-Book. By E. DE VERE BUCKINGHAM. Crown 8vo. 800 pages. Price 5s. It would be somewhat difficult to prove that a want exists in any direction; but, as a matter of fact, while there are price-books in all directions and, seemingly, of every kind, it has been found from practical experience that not one of these books is of much assistance to large contractors—by large contractors, we refer to such as are engaged in railway, dock, sewage, canal, water, and other works. This book is compiled to supply their needs, and will be published, corrected to date, annually.

The Construction of Carriageways and Footways By H. P. BOULNOIS, M.I.C.E., Past-President of Municipal and County Engineers, City Engineer of Liverpool. Demy 8vo. Illustrated. Price 5s.

Sewerage and Sewage Disposal of a Small Town. By E. B. SAVAGE, A.M.I.C.E. Demy 8vo. Illustrated. Price 5s

Management of Highways. By E. PURNELL HOOLEY, A.M.I.C.E., County Surveyor of Nottinghamshire. Demy 8vo. Price 1s.

IN THE PRESS.

Water Supply in Rural Districts. By R. GODFREY, A.M.Inst.C.E., Surveyor to the Rural Sanitary Authority, King's Lynn. Demy 8vo. Illustrated. Price 5s.

House Drainage. By W. SPINKS, A.M.Inst.C.E., Lecturer on Sanitary Engineering, Yorkshire College, Victoria University.

Designs and Discharging Capacities of Sewers, With Rules and Tables. By SANTO CRIMP, M.Inst.C.E., author of "Sewage Disposal."

Highway Bridges By E. P. SILCOCK, A.M.Inst.C.E., Borough Surveyor, King's Lynn.

Street and Town Sanitation. By C. MASON, M.Inst. C.E., Surveyor, St. Martin's-in-the-Fields, London, W.C.

www.ingramcontent.com/pod-product-compliance
Lightning Source LLC
Chambersburg PA
CBHW030052170426
43197CB00010B/1491